Lecture Notes of the Institute for Computer Sciences, Social Informatics and Telecommunications Engineering 277

More information about this series at http://www.springer.com/series/8197

Konstantin Avrachenkov · Longbo Huang ·
Jason R. Marden · Marceau Coupechoux ·
Anastasios Giovanidis (Eds.)

Game Theory
for Networks

8th International EAI Conference, GameNets 2019
Paris, France, April 25–26, 2019
Proceedings

Editors
Konstantin Avrachenkov
Inria Sophia Antipolis
Sophia Antipolis Cedex, France

Longbo Huang
Tsinghua University
Beijing, China

Jason R. Marden
University of California
Santa Barbara, CA, USA

Marceau Coupechoux
Telecom ParisTech (ENST)
Paris Cedex, France

Anastasios Giovanidis
Sorbonne Université
Paris Cedex, France

ISSN 1867-8211 ISSN 1867-822X (electronic)
Lecture Notes of the Institute for Computer Sciences, Social Informatics
and Telecommunications Engineering
ISBN 978-3-030-16988-6 ISBN 978-3-030-16989-3 (eBook)
https://doi.org/10.1007/978-3-030-16989-3

Library of Congress Control Number: 2019936141

This Springer imprint is published by the registered company Springer Nature Switzerland AG
The registered company address is: Gewerbestrasse 11, 6330 Cham, Switzerland

Preface

Game theory faces new challenges pertaining to the analysis and design of many emerging systems where networks play an increasingly prominent role, e.g., the Internet of Things, social networks, and machine learning. For example, game theory will be called to play a prominent role both in the design of new generation wireless networks (beyond 5G), and in the understanding of new complex interactions among social and economic agents in a digital world.

This book consists of peer-reviewed technical papers presented at GAMENETS 2019. They treat game theoretic applications related to wireless networks, economy and resource allocation, as well as social networks. We believe that these papers contribute toward advancing the fundamentals of game theory.

March 2019

Konstantin Avrachenkov
Longbo Huang
Jason R. Marden
Marceau Coupechoux
Anastasios Giovanidis

Organization

Steering Committee

Chair

Imrich Chlamtac   University of Trento, Italy

Member

Eitan Altman   Inria, France

Organizing Committee

General Co-chairs

Marceau Coupechoux   Telecom ParisTech, France
Anastasios Giovanidis   CNRS, France

TPC Chairs

Konstantin Avrachenkov   Inria, France
Longbo Huang   Tsinghua University, China
Jason R. Marden   University of California Santa Barbara, USA

Local Chair

Pierre Coucheney   UVSQ, France

Workshops Chair

Bruno Gaujal   Inria, France

Publicity and Social Media Chair

Ju Bin Song   Kyung Hee University, South Korea

Publications Chair

Mikael Touati   Orange Labs, France

Web Chair

Jonathan Krolikowski   CentraleSupélec, France

Technical Program Committee

Alonso Silva Allende	Safran, France
Philip Brown	University of Colorado (Colorado Springs), USA
Pierre Coucheney	Université de Versailles Saint-Quentin-en-Yvelines, France
Ceyhun Eksin	Texas A&M, USA
Zhixuan Fang	The Chinese University of Hong Kong, SAR China
Andrey Garnaev	WINLAB/Rutgers University, USA
Xiaowen Gong	Auburn University, USA
Yezekael Hayel	UAPV, France
Vijay Kamble	University of Illinois at Chicago, USA
David Leslie	Lancaster University and PROWLER.io, UK
Lorenzo Maggi	Nokia Bell Labs, France
Ramtin Pedarsani	University of California Santa Barbara, USA
H. Vincent Poor	Princeton University, USA
Jorge Poveda	University of Colorado (Boulder), USA
Bary Pradelski	ETH Zurich, Switzerland
Lillian Ratliff	University of Washington, USA
Behrouz Touri	University of California San Diego, USA
Haoran Yu	The Chinese University of Hong Kong, SAR China
Quanyan Zhu	New York University, USA

Contents

Game Theory for Wireless Networks

Jamming in Multiple Independent Gaussian Channels as a Game

Michail Fasoulakis[1(✉)], Apostolos Traganitis[1,2], and Anthony Ephremides[3]

[1] Institute of Computer Science, Foundation for Research and Technology-Hellas (ICS-FORTH), Heraklion, Greece
{mfasoul,tragani}@ics.forth.gr
[2] Department of Computer Science, University of Crete, Heraklion, Greece
[3] Department of Electrical and Computer Engineering,
Institute for Systems Research, University of Maryland,
College Park, USA
etony@umd.edu

Abstract. We study the problem of *jamming* in multiple independent *Gaussian channels* as a *zero-sum game*. We show that in the unique *Nash equilibrium* of the game the *best-response strategy* of the transmitter is the *waterfilling* to the sum of the jamming and the noise power in each channel and the *best-response strategy* of the jammer is the *waterfilling* only to the noise power.

Keywords: Wireless communications · Jamming · Zero-sum game · Gaussian channels

1 Introduction

One of the fundamental problems in reliability of wireless communications is the problem of *jamming*. This refers to the existence of a malicious user, the jammer, which tries to compromise or destroy the performance of the wireless link. We can view the problem of *jamming* as a *non-cooperative game* [7]. Several studies in the past studied this problem as a two-player game between the transmitter and the jammer. In particular, in [1] Altman et al. considered a *non zero-sum non-cooperative game* taking into account the transmission cost. More specifically, in this work the transmitter wants to maximize her own rate and the jammer wants to minimize the rate of the transmitter taking into account their transmission costs. The result of that work is the proof of the existence and the characterization of a unique *Nash equilibrium* and also an algorithm to compute it. In [5], the author studied the problem as a *zero-sum game* using as utility of

The research leading to these results has partially received funding from the Marie Sklodowska-Curie Actions - Initial Training Networks (ITN) European Industrial Doctorates (EID) project WiVi-2020 (H2020 MSCA-ITN, project no. 642743). For this work the first author is supported by the Stavros Niarchos Foundation-FORTH postdoc fellowship for the project ARCHERS.

K. Avrachenkov et al. (Eds.): GameNets 2019, LNICST 277, pp. 3–8, 2019.
https://doi.org/10.1007/978-3-030-16989-3_1

the transmitter the *linearized Shannon capacity*. In [2], the authors studied the problem in the case of multiple jammers. In [6] the authors studied the *jamming* problem of Gaussian MIMO channels as a *zero-sum game*.

Inspired by the work of [1], we study the problem of *jamming* as a *zero-sum game*, with one transmitter and one jammer over multiple independent *Gaussian channels* with perfect channel state information (CSI). Our scenario is a special case of the model in [1] when the transmission costs are zero and their results can be applied in our scenario as a subcase of their general model. Nevertheless, we believe that this scenario is of practical interest and in our work we analyze it in more detail with different and simpler tools giving simple and intuitively satisfying insights into the problem. In particular, we show that in the *Nash equilibrium* of the game the *best-response strategy* of the transmitter is to apply the well known *Waterfilling Theorem* of Information theory (see page 235 of [3]) taking into account the sum of the jamming plus noise power in each channel, and the *best-response strategy* of the jammer is to apply the *Waterfilling Theorem* taking into account only the noise power.

2   The Model

We consider the *Gaussian channel*, with one pair of transmitter-receiver and one malicious user, the jammer. There are $M > 0$ independent channels that the transmitter can use to transmit her information. In particular, the transmitter has a positive budget of transmission power T and wants to distribute it on the channels in a way that maximizes her aggregate rate. Let the non-negative T_k be the portion of the power T that is used in the channel k, with $\sum_{k=1}^{M} T_k = T$. In the absence of a jammer her optimum strategy is the well known *waterfilling strategy* with respect to the channel's noise. On the other hand, the jammer has a positive budget of transmission power J and wants to distribute it on the channels in a way that minimizes the aggregate rate of the transmitter. Let the non-negative J_k be the portion of the power J that is used in the channel k, with $\sum_{k=1}^{M} J_k = J$. Let $\alpha_T > 0$ be the channel attenuation for the transmitter-receiver pair equal for all channels and let $\alpha_J > 0$ be the channel attenuation for the jammer-receiver pair also equal for all channels. Also, let $N_k > 0$ be the power of the *additive Gaussian white noise (AGWN)* in the channel k. The receiver treats the signal of the jammer as noise, so by the *Shannon's formula* [4,8] the rate/utility of the transmitter in nats per channel use is:

$$u_T = R_T = \frac{1}{2} \sum_{k=1}^{M} \ln \left(1 + \frac{\alpha_T T_k}{\alpha_J J_k + N_k} \right),$$

which must be maximized.

On the other hand, the utility of the jammer is:

$$u_J = u_T,$$

which must be minimized.

We can consider the transmitter and the jammer as the players in a *zero-sum game* [7]. The strategies of the transmitter are the constants T_k that are used to distribute her power on the M channels and the strategies of the jammer are the constants J_k that are used to distribute her power on the M channels. Since, u_T is concave in T_k and convex in J_k, we can apply the *Sion's minimax Theorem* to conclude that it has a saddle point.

3 The Strategy of the Transmitter: Waterfilling

We will analyse the *best-response strategy* of the transmitter in the *zero-sum game* when the strategy of the jammer is fixed, in other words the parameters J_k are fixed. We have the following optimization problem:

$$\max_{T_k} \frac{1}{2} \sum_{k=1}^{M} \ln \left(1 + \frac{\alpha_T T_k}{\alpha_J J_k + N_k} \right)$$

$$= \max_{T_k} \frac{1}{2} \sum_{k=1}^{M} \ln \left(\alpha_J J_k + N_k + \alpha_T T_k \right)$$

$$- \frac{1}{2} \sum_{k=1}^{M} \ln \left(\alpha_J J_k + N_k \right)$$

$$\text{s.t.} \sum_{k=1}^{M} T_k = T,$$

$$T_k \geq 0, \qquad\qquad 1 \leq k \leq M.$$

We can see that only $\frac{1}{2} \sum_{k=1}^{M} \ln \left(\alpha_J J_k + N_k + \alpha_T T_k \right)$ depends on T_k and therefore the solution of this convex optimization problem is the well known *waterfilling theorem* of the Information Theory (see page 245 of [3]) which states that $\alpha_T T_k = (v - \alpha_J J_k - N_k)^+$, where v is calculated by the expression $\sum_{k=1}^{M} (v - \alpha_J J_k - N_k)^+ = \alpha_T T$. It is easy to see that $v > 0$. If for some channel(s) j, $\alpha_J J_j + N_j \geq v$ then $T_j = 0$ and the transmitter will apply the *waterfilling strategy* to the rest of the channels. This situation will arise only when there is excessive noise in some channels, that is if $N_j \geq v$, since the jammer will avoid wasting power in a channel which will not be used by the transmitter.

The *waterfilling strategy* of the transmitter maximizes her rate for any strategy of the jammer including the best one which minimizes this maximum.

4 The Strategy of the Jammer

The *best-response strategy* of the jammer in the *zero-sum game* if the strategy of the transmitter is fixed, that is if the powers T_k have specific values, is determined by the following optimization problem: Minimize u_T, that is

$$\min_{J_k} \frac{1}{2} \sum_{k=1}^{M} \ln \left(1 + \frac{\alpha_T T_k}{\alpha_J J_k + N_k} \right)$$

$$= \min_{J_k} \frac{1}{2} \sum_{k=1}^{M} \ln \left(\alpha_T T_k + \alpha_J J_k + N_k \right)$$

$$- \frac{1}{2} \sum_{k=1}^{M} \ln \left(\alpha_J J_k + N_k \right)$$

s.t.

$$\sum_{k=1}^{M} J_k = J,$$

$$J_k \geq 0, \qquad\qquad 1 \leq k \leq M.$$

For the analysis, we use the KKT conditions. We use a multiplier u for the equation $\sum_{k=1}^{M} J_k = J$ and a multiplier λ_k for any condition $J_k \geq 0$. Thus, by the KKT conditions we have, for any k, the condition

$$\frac{\alpha_J}{2(\alpha_T T_k + \alpha_J J_k + N_k)} - \frac{\alpha_J}{2(\alpha_J J_k + N_k)} + u = \lambda_k, \qquad (1)$$

the *complementarity slackness condition*

$$\lambda_k J_k = 0, \qquad (2)$$

and

$$\lambda_k \geq 0.$$

From the KKT conditions when J_k is positive, so $\lambda_k = 0$, must satisfy the condition $1/(\alpha_J J_k + N_k) - 1/(\alpha_T T_k + \alpha_J J_k + N_k) = \frac{2u}{\alpha_J}$, where $\frac{2u}{\alpha_J}$ is a positive constant, since it is easy to see that $u > 0$. Solving for J_k we find

$$J_k = \frac{1}{2\alpha_J} \left[- \alpha_T T_k - 2N_k + \sqrt{(\alpha_T T_k)^2 + 2\frac{\alpha_J \alpha_T T_k}{u}} \right]^+$$

and u is calculated by inserting the J_ks into equation $\sum_{k=1}^{M} J_k = J$. We can see that J_k increases as T_k increases. Thus this strategy of the jammer will convert an "attractive" (= less noisy) channel in which the transmitter applies larger power, into a less attractive (=more noisy) channel forcing the transmitter to apply less power (assuming that she follows a *waterfilling strategy*) and vice versa. This *best-response strategy* of the jammer is also a mechanism which can be used to force the behaviour of the transmitter in a manner that leads to the *Nash equilibrium* of the game which is derived in the next section.

5 The Nash Equilibrium Strategies

In this section, we will extend the previous results and observations to determine the strategies at the *Nash equilibrium*. In the *Nash equilibrium* the transmitter plays a *waterfilling strategy* by keeping constant the sum of her power, the power of the jammer and the noise power in each channel. The strategy of the jammer taking into account the optimal transmitter's strategy is determined by the *minimax optimization problem*, that is

$$\min_{J_k} \max_{T_k} \frac{1}{2} \sum_{k=1}^{M} \ln \left(1 + \frac{\alpha_T T_k}{\alpha_J J_k + N_k} \right)$$

$$= \min_{\substack{J_k \\ \text{Transmitter plays} \\ \text{waterfilling}}} \frac{1}{2} \sum_{k=1}^{M} \ln \left(1 + \frac{\alpha_T T_k}{\alpha_J J_k + N_k} \right)$$

$$= \min_{\substack{J_k \\ \text{Transmitter plays} \\ \text{waterfilling}}} \frac{1}{2} \sum_{k=1}^{M} \ln \left(\alpha_T T_k + \alpha_J J_k + N_k \right)$$

$$- \frac{1}{2} \sum_{k=1}^{M} \ln \left(\alpha_J J_k + N_k \right)$$

s.t.

$$\sum_{k=1}^{M} J_k = J,$$

$$J_k \geq 0, \qquad\qquad 1 \leq k \leq M.$$

The KKT conditions that we described in the previous section must hold at the *Nash equilibrium* of the game. For the analysis, we categorize the noise of a channel k into three groups according to its power, $N_k \geq v$, $N_k \in (\frac{v\alpha_J}{2uv+\alpha_J}, v)$ and $N_k \leq \frac{v\alpha_J}{2uv+\alpha_J}$.

By the *waterfilling strategy* of the transmitter as we analyse in Sect. 3, if $N_k \geq v$, then $T_k = 0$. Also, by the condition (1) we can see that if $T_k = 0$, then $\lambda_k > 0$, so by (2) we conclude that $J_k = 0$. Thus in the channels with excessive noise both the transmitter and the jammer will avoid wasting any power.

For $N_k \in (\frac{v\alpha_J}{2uv+\alpha_J}, v)$, there are two cases for the jamming power: $\alpha_J J_k \geq v - N_k$ and $\alpha_J J_k < v - N_k$. The first case can not happen since then $T_k = 0 \Rightarrow \lambda_k > 0 \Rightarrow J_k = 0$, but this is a contradiction. In the second case we have $\alpha_T T_k + \alpha_J J_k + N_k = v$ and $\lambda_k > 0$, since $\lambda_k = \frac{\alpha_J}{2v} - \frac{\alpha_J}{2(\alpha_J J_k + N_k)} + u \geq \frac{\alpha_J}{2v} - \frac{\alpha_J}{2N_k} + u > \frac{\alpha_J}{2v} - \frac{(2uv+\alpha_J)}{2v} + u = u - u = 0 \Rightarrow J_k = 0$. Thus the channels with $N_k \in (\frac{v\alpha_J}{2uv+\alpha_J}, v)$ will be used by the transmitter but not by the jammer since the noise is large enough to make jamming inefficient and a waste of jamming power.

For $N_k \leq \frac{v\alpha_J}{2uv+\alpha_J}$, the only possibility for the jamming power is $\alpha_J J_k < v - N_k$ (for the same reason as above). In this case we have $\alpha_T T_k + \alpha_J J_k + N_k = v$ and $\lambda_k = 0$, since the case $\lambda_k > 0 \Rightarrow J_k = 0$ leads to the contradiction $N_k > \frac{v\alpha_J}{2uv+\alpha_J}$. Setting $\alpha_T T_k + \alpha_J J_k + N_k = v$ and $\lambda_k = 0$ in (1) and solving for J_k we find that $\alpha_J J_k = \frac{v\alpha_J}{\alpha_J+2uv} - N_k$. The constant u can be found by solving the equation $\sum_{k=1}^{M} (\frac{v\alpha_J}{\alpha_J+2uv} - N_k)^+ = \alpha_J J$. The preceding analysis proves that the strategy of the jammer in the *Nash equilibrium* is also the *waterfilling* with respect to the noise power of the channels.

In particular, if the powers of the transmitter and the jammer are much larger than the noise power then the *waterfilling strategy* of the jammer makes the combined jamming plus noise power equal in all channels whereas the *waterfilling strategy* of the transmitter results in the uniform distribution of her power in all channels.

6   Conclusions

In this paper we study the problem of *jamming* in multiple independent *Gaussian channels*. We derive the strategies of the transmitter and the jammer on the unique *Nash equilibrium* where the transmitter maximizes the minimum of her rate and the jammer minimizes the maximum of the transmitter rate. In particular, given that the *waterfilling strategy* of the transmitter is known as the best strategy under interference, our main contribution in this paper is to show that the jammer's optimum strategy in the *Nash equilibrium* is the *waterfilling* to the noise of the channels as well.

References

1. Altman, E., Avrachenkov, K., Garnaev, A.: A jamming game in wireless networks with transmission cost. In: Chahed, T., Tuffin, B. (eds.) NET-COOP 2007. LNCS, vol. 4465, pp. 1–12. Springer, Heidelberg (2007). https://doi.org/10.1007/978-3-540-72709-5_1
2. Altman, E., Avrachenkov, K., Garnaev, A.: Jamming in wireless networks: the case of several jammers. In: Proceedings of the 1st International Conference on Game Theory for Networks (GAMENETS 2009), pp. 585–592 (2009)
3. Boyd, S., Vandenberghe, L.: Convex Optimization. Cambridge University Press, Cambridge (2004)
4. Cover, T.M., Thomas, J.A.: Elements of Information Theory. Wiley, London (2006)
5. Garnaev, A.: Modeling Computation and Optimization, Chap. 1. World Scientific Publishing Company Incorporated, Singapore (2009)
6. Jorswieck, E.A., Boche, H., Weckerle, M.: Optimal transmitter and jamming strategies in Gaussian MIMO channels. In: Vehicular Technology Conference (VTC 2005), vol. 2, pp. 978–982 (2005)
7. Osborne, M.J., Rubinstein, A.: A Course in Game Theory. MIT Press, Cambridge (1994)
8. Shannon, C.E.: A mathematical theory of communication. Bell Syst. Tech. J. **27**(4), 623–656 (1948)

Bandwidth Scanning when the Rivals Are Subjective

Andrey Garnaev$^{(\boxtimes)}$ and Wade Trappe

WINLAB, Rutgers University, North Brunswick, USA
garnaev@yahoo.com, trappe@winlab.rutgers.edu

Abstract. In this paper we consider how subjectivity affects the problem of scanning spectrum bands, and the impact on both the scanner and invader's strategy. To model such subjective behavior, we formulate a prospect theoretical (PT) extension of the Bayesian bandwidth scanning game where the Scanner knows only a priori probabilities about what type of intrusion (e.g. regular intensity or low intensity) occurs in the spectrum bands. Existence and uniqueness of the PT Bayesian equilibrium is proven. Moreover, these PT Bayesian equilibrium strategies are derived in closed form as functions of the detection probabilities associated with different invader types. Waterfilling equations are derived, which allows one to determine these detection probabilities. Bands where the Invader's strategies have band-sharing form are identified. The sensitivity of the strategies to the subjective factors and a priori probabilities are numerically illustrated.

Keywords: Bayesian equilibrium · Prospect theory ·
Bandwidth scanning

1 Introduction

Cognitive radio networks will support dynamic spectrum access (DSA). However, in spite of the potential benefits for DSA, the open nature of the wireless medium will make cognitive radios a powerful tool for conducting malicious activities or policy violations by secondary users. Therefore, detecting malicious users or unlicensed activities is a crucial problem facing DSA [19], and one of the challenges to enforcing the proper usage of spectrum is the development of an intrusion detection system (IDS) that will scan large amounts of spectrum and identify anomalous activities [4,6,19]. Since, in such security problems, there are two agents with different goals (the IDS aims to detect illegal spectrum usage, while the adversary intends to sneak into bands undetected for their illegal usage), *game theory* is an ideal tool to employ [13]. As examples of applying game theory to detect an adversary to prevent malicious attack on networks, we mention [2,3,5,8,9,11,12,14–16,20,23,25,27–30]. In all of these papers the rivals were *rational*.

K. Avrachenkov et al. (Eds.): GameNets 2019, LNICST 277, pp. 9–28, 2019.
https://doi.org/10.1007/978-3-030-16989-3_2

Prospect Theory (PT) [17] has been developed to model the *subjective* factor in an agent's behavior. In particular, PT models agents with subjective behavior by employing subjective probabilities, rather than the objective probabilities that might be used in rational behavior, to weight the values of possible outcomes. We note that, although PT originally was designed to take into account the possibility of the risk of irrational behavior by rivals in economic problems [17,18], it has been applied recently to different network and communication security problems. For example, in [24], for designing Trojan detection algorithm, in [31,32], for developing anti-jamming strategies in cognitive radio networks, in [26], for designing secure drone delivery systems, and in [33], for maintaining a cloud storage defense against advanced persistent threats.

In this paper, we consider how subjectivity affects the problem of scanning large amounts of bandwidth to detect illegal intrusion, where the scanner has incomplete information about the Invader's characteristics. To gain insight into this problem, we formulate a *PT extension* of the Bayesian scanning game between a *Scanner* and an *Invader* considered in [10]. We prove that this PT extension has a unique solution, and find the PT equilibrium strategies in closed form. To the best knowledge of the authors, this paper presents the first PT equilibrium strategies in closed form for an n-dimension payoff matrix and any subjective probabilities. The closed form solution allows us to reveal some interesting properties of the solution, such as the water-filling structure of the PT strategies as well as to identify the bands where Invader's strategies have band-sharing form. The solution also allows one to observe the sensitivity of the strategies on the subjective factors and on a priori information about the Invader's type.

The organization of this paper is as follows. In Sect. 2, the basic model for bandwidth scanning as a zero sum with a diagonal payoff matrix is formulated. In Sect. 3, the basics of prospect theory are presented as a basis for the rest of the paper. In Sect. 4, the PT extension of the basic bandwidth scanning game is described and solved. Next, in Sect. 5, the Bayesian extension of the basic bandwidth scanning game is formulated for the case where there is incomplete information about the mode of intrusion attack. In Sect. 7, auxiliary assumptions, notations and results are introduced, and they are employed in Sect. 8 to derive the equilibrium strategies for the PT Bayesian scanning game. Finally, in Sect. 9, conclusions are offered.

2   Model

In this section, we describe our basic problem model, which involves a scenario where a primary user (i.e. the Scanner) owns n frequency bands $1, 2, \ldots, n$, which will be scanned by the Scanner. The Invader will attempt to "sneak" usage on *only* one of these bands, while the Scanner can only scan a single band at a time. We assume that the Invader will be detected with probability γ_i if he sneaks in band i and the Scanner scans that band. We note that the detection probability of the Invader depends on its SINR [21], and thus, in particular, on the power of the Invader's signal as well as on the distance between the Invader and the Scanner. If the Scanner does not scan the band that the Invader is using, then the Invader sneaks safely, i.e., its detection probability is zero. We assume that the payoff to the Scanner is one if the Invader is detected and it is zero otherwise. Let a (mixed) strategy for the Scanner be $\boldsymbol{x} = (x_1, \ldots, x_n)$, where x_i is the probability (reflecting the likelihood of revisiting that channel when the game is repeated) that he scans band i. So, $\sum_{i=1}^{n} x_i = 1$ and $x_i \geq 0$, $i = 1, \ldots, n$. Let a (mixed) strategy for the Invader be $\boldsymbol{y} = (y_1, \ldots, y_n)$, where y_i is the probability that he sneaks in band i. Thus, $\sum_{i=1}^{n} y_i = 1$ and $y_i \geq 0$, $i = 1, \ldots, n$. Denote by $\mathcal{P}$, the set of all n dimensional probability vectors. Thus, $\mathcal{P}$ is the set of all feasible strategies for the Scanner as well as for the Invader. Then, the expected payoff to the Scanner if the rivals employ strategy $\boldsymbol{x}$ and $\boldsymbol{y}$ is given as follows:

$$V(\boldsymbol{x}, \boldsymbol{y}) = \sum_{i=1}^{n} \gamma_i x_i y_i. \tag{1}$$

This payoff reflects detection probability of the Invader. The Scanner wants to maximize this detection probability, while the Invader wants to minimize it. Thus, this is a zero sum game, and we look for its equilibrium. Recall that $(\boldsymbol{x}, \boldsymbol{y})$ is an equilibrium if and only if for any $(\tilde{\boldsymbol{x}}, \tilde{\boldsymbol{y}}) \in \mathcal{P} \times \mathcal{P}$ the following inequalities hold:

$$V(\tilde{\boldsymbol{x}}, \boldsymbol{y}) \leq V(\boldsymbol{x}, \boldsymbol{y}) \leq V(\boldsymbol{x}, \tilde{\boldsymbol{y}}) \tag{2}$$

This implies that $\boldsymbol{x}$ and $\boldsymbol{y}$ are equilibrium strategies if and only if they are the best response to each other, i.e., they are solutions of the following best response equations:

$$\boldsymbol{x} = \mathrm{BR}_S(\boldsymbol{y}) := \underset{\boldsymbol{x} \in \mathcal{P}}{\arg\max}\, V(\boldsymbol{x}, \boldsymbol{y}), \tag{3}$$

$$\boldsymbol{y} = \mathrm{BR}_I(\boldsymbol{x}) := \underset{\boldsymbol{y} \in \mathcal{P}}{\arg\min}\, V(\boldsymbol{x}, \boldsymbol{y}). \tag{4}$$

This is a zero sum game with diagonal payoff matrix and its equilibrium strategies will be given in closed form in Corollary 1 of Sect. 4.

3   Basics of Prospect Theory

In this section, we will briefly review the basic concepts associated with the PT solution. Prospect theory, which was introduced by Tversky and Kahneman [17], is a method for describing decisions under a *subjective* factor. In particular, it is revealed in PT that agents use subjective probabilities ("decision weight") $w(p)$, rather than the objective probabilities p, to weight the values of possible outcomes. Moreover, agents tend to over-weight low probability outcomes and under-weight higher probability outcomes. This feature is captured by weighting the probability distribution by the *weighting function* $w(p)$. Such weighting functions satisfy four basic properties in the plane (p, w) with $p \in [0, 1]$ [22]: (a) regressive intersecting the diagonal from above, (b) asymmetric with fixed point at about $1/3$, (c) S-shaped concave on an initial interval and convex beyond that, and (d) reflective assigning equal weight to a given loss-probability as to a given gain-probability. Although all of our obtained analytical results hold for any weighting function, as a basic example of the probability weighting function we consider the Prelec function in our numerical examples, which is as follows [22]:

$$w(p) := e^{-(-\ln(p))^{\kappa}} \tag{5}$$

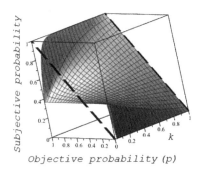

Fig. 1. Subjective probability as function on objective probability p and probability weighting parameter κ.

with $\kappa \in (0, 1]$ is the *probability weighting parameter*. The probability weighting parameter reflects the distortion from the true objective probability that is caused by the subjective evaluation of the agent. In other words, this parameter allows us to measure how rational (or subjective) an agent is. Namely, $\kappa = 1$ corresponds to rational agent, while smaller κ corresponds to a less rational agent, or in other words, a more subjective agent. Figure 1 illustrates the dependence of the subjective probabilities given by (5) on the objective probability p and the probability weighting parameter κ. For boundary case $\kappa \downarrow 0$, the subjective probability $w(p)$ approximates a step function, flat everywhere except near the endpoints of the probability interval. While, for the other boundary case of $\kappa = 1$, this subjective probability $w(p)$ coincides with objective probability, i.e.,

$$w(p) \equiv p \text{ for } \kappa = 1, \tag{6}$$

i.e., $\kappa = 1$ corresponds to full rationality of the agent. Finally, in particular, $w(p)$ has the following monotonicity property:

$$w(p) \text{ is strictly increasing from 0 for } p = 0 \text{ to 1 for } p = 1. \tag{7}$$

4   Subjective Rivals

Denote by $w_S(p)$ and $w_I(p)$ the probability weighting functions employed by the Scanner and the Invader correspondingly. Then, the PT-utilities for the rivals associated with the zero sum game with payoff (21) are given as follows:

$$u_S^{PT}(\boldsymbol{x}, \boldsymbol{y}) := \sum_{i=1}^{n} \gamma_i x_i w_S(y_i) \text{ and } u_I^{PT}(\boldsymbol{x}, \boldsymbol{y}) := \sum_{i=1}^{n} \gamma_i w_I(x_i) y_i. \tag{8}$$

Then, under PT equilibrium, we consider the solution of the following PT best response equations

$$\boldsymbol{x} = \mathrm{BR}_S^{PT}(\boldsymbol{y}) := \underset{\boldsymbol{x} \in \mathcal{P}}{\mathrm{argmax}}\, u_S^{PT}(\boldsymbol{x}, \boldsymbol{y}), \tag{9}$$

$$\boldsymbol{y} = \mathrm{BR}_I^{PT}(\boldsymbol{x}) := \underset{\boldsymbol{y} \in \mathcal{P}}{\mathrm{argmin}}\, u_I^{PT}(\boldsymbol{x}, \boldsymbol{y}). \tag{10}$$

Theorem 1. *The game has the unique PT equilibrium* $(\boldsymbol{x}, \boldsymbol{y})$, *where*

$$x_i = x_i(\omega) = w_I^{-1}\left(\frac{\omega}{\gamma_i}\right) \text{ and } y_i = y_i(\nu) = w_S^{-1}\left(\frac{\nu}{\gamma_i}\right), \tag{11}$$

where $\omega \in (0, \min \underline{\gamma})$ *and* $\nu \in (0, \min \underline{\gamma})$ *with* $\underline{\gamma} = \min_i \gamma_i$ *uniquely defined as solutions of the equations:*

$$\sum_{i=1}^{n} w_I^{-1}\left(\frac{\omega}{\gamma_i}\right) \text{ and } \sum_{i=1}^{n} w_S^{-1}\left(\frac{\nu}{\gamma_i}\right) = 1. \tag{12}$$

Proof. Since $u_S^{PT}(\boldsymbol{x}, \boldsymbol{y})$ is linear in $\boldsymbol{x}$, $\boldsymbol{x}$ is the best response strategy to $\boldsymbol{y}$ if and only if there is a ν such that

$$\gamma_i w_S(y_i) \begin{cases} = \nu, & x_i > 0, \\ \leq \nu, & x_i = 0. \end{cases} \tag{13}$$

Since $u_I^{PT}(\boldsymbol{x}, \boldsymbol{y})$ is linear on $\boldsymbol{y}$, $\boldsymbol{y}$ is the best response strategy to $\boldsymbol{x}$ if and only if there is ω such that

$$\gamma_i w_I(x_i) \begin{cases} = \omega, & y_i > 0, \\ \geq \omega, & y_i = 0. \end{cases} \tag{14}$$

The assumption that there is an i such that $x_i = 0$ leads to a contradiction, since, by (14), $\omega = 0$, and so $x_i = 0$ for any i. Thus,

$$x_i > 0 \text{ for any } i. \tag{15}$$

Assume, now that there is an i such that $y_i = 0$. Then, by (13), $x_i = 0$. This contradicts (15). Thus, also $y_i > 0$ for any i. Thus, by (13) and (14),

$$\gamma_i w_S(y_i) = \nu \text{ and } \gamma_i w_I(x_i) = \omega \text{ for any } i. \tag{16}$$

This and (17) imply (11). Finally (17), (11) and the fact that $\boldsymbol{x}$ and $\boldsymbol{y}$ are probability vectors yield (12). ∎

Remark 1. *The inverse of the weighting function (5) is given as follows:*

$$w^{-1}(p) := e^{-(-\ln(p))^{1/\kappa}}. \tag{17}$$

It is apparent that the scenario with rational rivals is a boundary case of the subjective behavior with $\kappa = 1$, and it corresponds to zero-sum game with diagonal payoff matrix [7]. Namely, for rational rivals the following result holds.

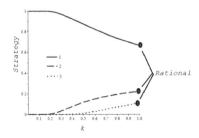

Fig. 2. PT equilibrium strategy as a function of the probability weighting parameter κ.

Corollary 1. *In the game with a rational Scanner and Invader, the equilibrium strategies (x, y) are unique and given as follows:*

$$x_i = y_i = \frac{1/\gamma_i}{\sum_{j=1}^{n}(1/\gamma_j)} \text{ for } i = 1, \dots, n. \tag{18}$$

Proof: For rational rivals $w_I(p) \equiv w_S(p) \equiv p$. Thus, by (11), $x_i(\omega) = \omega/\gamma_i$ and $y_i(\nu) = \nu/\gamma_i$. Then, by (12), $\omega = \nu = 1/\sum_{j=1}^{n}(1/\gamma_j)$, and the result follows. ∎

Thus, if both rivals are rational their equilibrium strategies coincide and both of them equalize the detection probability at each band. Meanwhile, the PT equilibrium strategies can differ if the subjective factors for the rivals differ, and the strategies are then given by water-filling equations (12).

Let us illustrate dependence of PT equilibrium strategies on κ by an example with $n = 3$ bands and detection probabilities $\gamma = (0.1, 0.3, 0.6)$. Figure 2 illustrates that the more subjective agent (which is reflected by smaller probability weighting parameter κ) focuses more effort (scanning for the Scanner and intrusion for the Invader) on the band with the smallest detection probability, i.e., band 1, and reduces its effort on all of the other bands. On the other hand, the more objective agent would spread its effort amongst the bands in such a way to equalize the detection probability when it becomes to be rational. An important property of these strategies is that each band will be scanned with positive probability, and likewise any band may be employed to sneak in with positive probability although these probabilities could approximately be zero when the probability weighting parameter tends to zero. As a point of reference, one can observe similar strategy behavior in the design of α-fair solutions as one increases the fairness coefficient [1].

5   Bayesian Game with Rational Rivals

In this section we return to the scenario involving rational rivals. Recall that the detection probability of the Invader depends on its SINR [21], and thus, in particular, on the power of the Invader's signal as well as on the distance between the Invader and Scanner. With each combination of transmitted signal by the Invader and its allocation, we can associate a type for the Invader and the corresponding detection probabilities in bands. In this paper, we will assume the Invader can be one of two types, an agent performing *regular intrusion* (denoted by type-1) and an agent performing *low intensity intrusion* (denoted by type-2). Let, with a priori probability α^t, a type-t Invader occurs, then $\alpha^1 + \alpha^2 = 1$. Let the detection probability be γ_i^t for the type-t Invader to be detected when it sneaks in band i and the Scanner scans this band. A (mixed) strategy for the type-t Invader is $\boldsymbol{y}^t = (y_1^t, \ldots, y_n^t)$, where y_i^t is the probability that he sneaks in band i. Thus, $\sum_{i=1}^n y_i^t = 1$ and $y_i^t \geq 0$, $i = 1, \ldots, n$. Let $\boldsymbol{Y} = (\boldsymbol{y}^1, \boldsymbol{y}^2)$. Then, we consider the probability of detection of the Invader if the rivals employ strategy $\boldsymbol{x}$ and $\boldsymbol{Y}$ as the expected payoff to the Scanner:

$$V(\boldsymbol{x}, \boldsymbol{Y}) = \sum_{i=1}^n \left(\alpha^1 \gamma_i^1 x_i y_i^1 + \alpha^2 \gamma_i^2 x_i y_i^2 \right), \tag{19}$$

while the probabilities to be detected are considered as cost functions of the corresponding Invader's type:

$$V^t(\boldsymbol{x}, \boldsymbol{y}^t) = \sum_{i=1}^n \gamma_i^t x_i y_i^t. \tag{20}$$

We look for Bayesian equilibrium, i.e., for such strategies $(\boldsymbol{x}, \boldsymbol{Y})$ that for any $(\tilde{\boldsymbol{x}}, \tilde{\boldsymbol{Y}})$ the following inequalities hold:

$$\begin{aligned} V(\tilde{\boldsymbol{x}}, \boldsymbol{Y}) &\leq V(\boldsymbol{x}, \boldsymbol{Y}), \\ V^t(\boldsymbol{x}, \boldsymbol{y}^t) &\leq V^t(\boldsymbol{x}, \tilde{\boldsymbol{y}}^t), t = 1, 2. \end{aligned} \tag{21}$$

This implies that $\boldsymbol{x}$ and $\boldsymbol{Y}$ are equilibrium strategies if and only if they are the best response to each other, i.e., they are solutions of the following best response equations:

$$\boldsymbol{x} = \mathrm{BR}_S(\boldsymbol{Y}) := \underset{\boldsymbol{x} \in \mathcal{P}}{\operatorname{argmax}} \, V(\boldsymbol{x}, \boldsymbol{Y}), \tag{22}$$

$$\boldsymbol{y}^t = \mathrm{BR}_I^t(\boldsymbol{x}) := \underset{\boldsymbol{y}^t \in \mathcal{P}}{\operatorname{argmin}} \, V^t(\boldsymbol{x}, \boldsymbol{y}^t) \text{ for } t = 1, 2. \tag{23}$$

Unique equilibrium strategies of this Bayesian game will be given in Corollary 2 of Sect. 8.

6   Bayesian Game with Subjective Rivals

In this section, we formulate the PT extension of the Bayesian game from the previous section. We now let the rivals be subjective. Thus, to define such PT extension, first, we have to define PT-utilities for the rivals associated with the Bayesian game with payoff (19) and cost functions (20) given as follows:

$$
\begin{aligned}
u_S^{PT}(\boldsymbol{x}, \boldsymbol{Y}) &:= \sum_{i=1}^{n} \left(\alpha^1 \gamma_i^1 x_i w_S(y_i^1) + \alpha^2 \gamma_i^2 x_i w_S(y_i^2) \right), \\
u_I^{PT,t}(\boldsymbol{x}, \boldsymbol{y}^t) &:= \sum_{i=1}^{n} \gamma_i^t w_I(x_i) y_i^t.
\end{aligned}
\tag{24}
$$

Then, under the PT equilibrium $(\boldsymbol{x}, \boldsymbol{Y})$ of the Bayesian game, we consider the solution of the following PT best response equations:

$$
\boldsymbol{x} = \mathrm{BR}_S^{PT}(\boldsymbol{Y}) := \operatorname*{argmax}_{\boldsymbol{x}} u_S^{PT}(\boldsymbol{x}, \boldsymbol{Y}),
\tag{25}
$$

$$
\boldsymbol{y}^t = \mathrm{BR}_I^{PT,t}(\boldsymbol{x}) := \operatorname*{argmin}_{\boldsymbol{y}^t} u_I^{PT,t}(\boldsymbol{x}, \boldsymbol{y}^t), t = 1, 2.
\tag{26}
$$

In Theorem 2 of Sect. 8, it will be proven that these PT best response equations have a unique solution and the PT equilibrium strategies will be found in closed form.

7   Auxiliary Assumption, Notations and Results

In this section, we introduce auxiliary notations and results. First, to avoid bulkiness in formulas, we assume that there are no two different bands with the same ratio of detection probabilities, i.e., $\gamma_i^2/\gamma_i^1 \neq \gamma_j^2/\gamma_j^1$ for $i \neq j$. Then, without loss in generalization we can assume that the bands are arranged in such an order that

$$
\gamma_1^2/\gamma_1^1 > \gamma_2^2/\gamma_2^1 > \ldots > \gamma_n^2/\gamma_n^1.
\tag{27}
$$

Let us now extend the definition of inverse functions $w_S^{-1}(\xi)$ and $w_J^{-1}(\xi)$ as follows:

$$
w_S^{-1}(\xi) := \begin{cases} w_S^{-1}(\xi), & \xi \leq 1, \\ 1, & \xi > 1 \end{cases} \quad \text{and} \quad w_I^{-1}(\xi) := \begin{cases} w_I^{-1}(\xi), & \xi \leq 1, \\ 1, & \xi > 1. \end{cases}
\tag{28}
$$

Let

$$
\underline{\Psi}_i(\nu) := \sum_{j=1}^{i-1} w_S^{-1}\left(\frac{\nu}{\alpha^1 \gamma_j^1} \right),
\tag{29}
$$

$$
\overline{\Psi}_i(\nu) := \sum_{j=i+1}^{n} w_S^{-1}\left(\frac{\nu}{\alpha^2 \gamma_j^2} \right),
\tag{30}
$$

and

$$
\underline{A}_i^1 := \min_{j \leq i-1} \alpha^1 \gamma_j^1 \quad \text{and} \quad \overline{A}_i^2 := \min_{j \geq i+1} \alpha^2 \gamma_j^2.
\tag{31}
$$

Proposition 1. (a) *For a fixed i, $\underline{\Psi}_i(\xi)$ is strictly increasing on ξ from zero for $\xi = 0$ to $\underline{\Psi}_i(\underline{A}_i^1) > 1$ for $\xi = \underline{A}_i^1$.*
(b) *For a fixed $\xi > 0$, $\underline{\Psi}_i(\xi)$ is increasing on i.*
(c) *For a fixed i, $\overline{\Psi}_i(\xi)$ is strictly increasing on ξ from zero for $\xi = 0$ to $\overline{\Psi}_i(\overline{A}_i^2) > 1$ for $\xi = \overline{A}_i^2$.*
(d) *For a fixed $\xi > 0$, $\overline{\Psi}_i(\xi)$ is decreasing on i.*

Proof: (a) and (c) follow from (17) and (28), while (b) and (d) follow from the following relations

$$\underline{\Psi}_{i+1}(\xi) - \underline{\Psi}_i(\xi) = w_S^{-1}\left(\xi/(\alpha^1 \gamma_i^1)\right) > 0,$$
$$\overline{\Psi}_{i+1}(\xi) - \overline{\Psi}_i(\xi) = -w_S^{-1}\left(\xi/(\alpha^2 \gamma_{i+1}^2)\right) < 0,$$

and the result follows. ∎

Based on Proposition 1, we can define two auxiliary finite sequences ν_i^1, ν_i^2, $i = 1, \ldots, n$ as follows:

$$\nu_i^1 = \begin{cases} \infty, & i = 1, \\ \text{the unique root in } (0, \underline{A}_i^1) \text{ of equation } \underline{\Psi}_i(\nu_i^1) = 1, & i = 2, \ldots, n \end{cases} \tag{32}$$

and

$$\nu_i^2 = \begin{cases} \text{the unique root in } (0, \overline{A}_i^2) \text{ of equation } \overline{\Psi}_i(\nu_i^2) = 1, & i = 1, \ldots, n-1, \\ \infty, & i = n. \end{cases} \tag{33}$$

These sequences have the following monotonicity properties:

Proposition 2. (a) $\nu_1^1 > \nu_2^1 > \ldots > \nu_{n-1}^1 > \nu_n^1$,
(b) $\nu_1^2 < \nu_2^2 < \ldots < \nu_{n-1}^1 < \nu_n^1$.

Proof: (a) follows from Proposition 1(a) and (b), while (b) follows from Proposition 1(c) and (d). ∎

Proposition 3. *There exists the unique $m_* \in \{1, \ldots, n\}$ such that*

$$\nu_j^1 \begin{cases} > \nu_j^2, & j < m_*, \\ \geq \nu_j^2, & j = m_*, \\ < \nu_j^2, & j > m_*. \end{cases} \tag{34}$$

Proof: Since $\nu_1^1 = \infty > \nu_1^2$ and $\nu_n^1 < \infty = \nu_n^2$, the result straightforward follows from Proposition 2. ∎

Proposition 4. *There is a unique i_* such that one of the following relations holds:*

$$\nu_{i_*}^1 = \nu_{i_*}^2, \tag{35}$$

$$\nu_{i_*-1}^2 \leq \nu_{i_*}^1 < \nu_{i_*}^2 \tag{36}$$

or

$$\nu_{i_*+1}^1 \leq \nu_{i_*}^2 < \nu_{i_*}^1. \tag{37}$$

Moreover,

$$i_* = \begin{cases} m_* + 1, & if\ (36)\ holds, \\ m_*, & if\ (37)\ or\ (35)\ holds. \end{cases} \tag{38}$$

Proof: First assume that $\nu_i^1 \neq \nu_i^2$ for any i. Thus, (35) cannot hold. Then, by Proposition 3,

$$\nu_i^2 < \nu_{m_*}^2 < \nu_{m_*}^1 < \nu_j^1 \text{ for any } i, j < m_* \tag{39}$$

and

$$\nu_j^1 < \nu_{m_*+1}^1 < \nu_{m_*+1}^2 < \nu_i^2 \text{ for any } i, j > m_* + 1. \tag{40}$$

Thus, neither (36) nor (37) can hold for any $i_* \notin \{m_*, m_* + 1\}$. So, if i_* exists then $i_* \in \{m_*, m_*+1\}$. Moreover, by (39), if (37) holds then $i_* = m_*$. While, by (40), if (36) holds then $i_* = m_* + 1$. This proves that if i_* exists then (38) has to hold. Thus, to complete the proof we have to prove that for i_* given by (38) only one of two relations (36) and (37) always holds. Let us consider separately two cases: (i) $\nu_{m_*+1}^1 \leq \nu_{m_*}^2 < \nu_{m_*}^1$ and (ii) $\nu_{m_*}^2 < \nu_{m_*}^1$ and $\nu_{m_*}^2 < \nu_{m_*+1}^1$. It is clear that (i) is equivalent to (37) with $i_* = m_*$. Let (ii) hold. Then, by (40), (ii) is equivalent to $\nu_{m_*}^2 < \nu_{m_*+1}^1 < \nu_{m_*+1}^2$, and this coincides with (36) with $i_* = m_* + 1$.

The case that there exists an i such that $\nu_i^1 = \nu_i^2$ can be considered similarly, and the result follows. ∎

8   Solution of the PT Bayesian Game

In this Section, we prove the uniqueness of the PT Bayesian equilibrium and find it in closed form. To prove the result we are going to employ a constructive approach. Namely, we first derive the necessary and sufficient conditions for the strategies $(\boldsymbol{x}, \boldsymbol{Y})$ to be the PT equilibrium, and establish what structure these strategies must have to satisfy these conditions. Then, taking into account that the strategies are probability vectors, we show that only one pair of strategies can have such a structure, and we design this pair of strategies in closed form.

Theorem 2. *The game has a unique Bayesian PT equilibrium $(\boldsymbol{x}, \boldsymbol{Y})$. Moreover, the Scanner's strategy $\boldsymbol{x}$ is given as follows*

$$x_j = x_j(\omega) = \begin{cases} w_I^{-1}\left(\dfrac{\omega}{\gamma_j^1}\right), & j \leq i_*, \\ w_I^{-1}\left(\dfrac{\gamma_{i_*}^2 \omega}{\gamma_{i_*}^1 \gamma_j^2}\right), & j \geq i_* + 1, \end{cases} \tag{41}$$

where i_* is given by Proposition 4, while ω is the unique root in $(0, \overline{\omega}_{i_*})$ of the equation

$$\sum_{j=1}^{i_*} w_I^{-1}\left(\frac{\omega}{\gamma_j^1}\right) + \sum_{j=i_*+1}^{n} w_I^{-1}\left(\frac{\gamma_{i_*}^2 \omega}{\gamma_{i_*}^1 \gamma_j^2}\right) = 1, \tag{42}$$

where $\overline{\omega}_i := \min\{\underline{\gamma}_i^1, \frac{1}{\gamma_i^2}\overline{\gamma}_i^2\}$) with $\underline{\gamma}_i^1 = \min_{j \leq i} \gamma_j^1$ and $\overline{\gamma}_i^2 = \min_{j > i} \gamma_j^2$.

This ω can be found by the bisection method since the left side of Eq. (42) is an increasing function of ω from zero for $\omega = 0$ and becomes greater than one for $\omega = \overline{\omega}$.

The Invader's strategy Y is given as follows:

$$y_i^1 = y_i^1(\nu) := \begin{cases} w_S^{-1}\left(\frac{\nu}{\alpha^1 \gamma_i^1}\right), & i \leq i_* - 1, \\ 1 - \sum_{j=1}^{i_*-1} w_S^{-1}\left(\frac{\nu}{\alpha^1 \gamma_j^1}\right), & i = i_*, \\ 0, & i \geq i_* + 1, \end{cases} \tag{43}$$

$$y_i^2 = y^2(\nu) := \begin{cases} 0, & i \leq i_* - 1, \\ 1 - \sum_{j=i_*+1}^{n} w_S^{-1}\left(\frac{\nu}{\alpha^2 \gamma_j^2}\right), & i = i_*, \\ w_S^{-1}\left(\frac{\nu}{\alpha^2 \gamma_i^2}\right), & i \geq i_* + 1, \end{cases} \tag{44}$$

where ν is the unique root in $(0, \underline{\nu}_{i_*}]$ with $\underline{\nu}_i := \min\{\nu_i^1, \nu_i^2\}$ of the equation

$$\Phi_{i_*}(\nu) = \nu \tag{45}$$

with

$$\Phi_i(\nu) = \alpha^1 \gamma_i^1 w_S \left(1 - \sum_{j=1}^{i-1} w_S^{-1}\left(\frac{\nu}{\alpha^1 \gamma_j^1}\right)\right)$$
$$+ \alpha^2 \gamma_i^2 w_S \left(1 - \sum_{j=i+1}^{n} w_S^{-1}\left(\frac{\nu}{\alpha^2 \gamma_j^2}\right)\right). \tag{46}$$

Due to function $\Phi_{i_*}(\nu)$ is decreasing on ν such that $\Phi_{i_*}(0) > 0$ for $\nu = 0$ and $\Phi_{i_*}(\underline{\nu}_{i*}) \leq \underline{\nu}_{i*}$ for $\nu = \underline{\nu}_{i*}$, the unique root of (45) in $(0, \underline{\nu}_{i*}]$ can be found by bisection method.

Finally, note that ν is the probability that the Invader is detected by the Scanner, ω is the detection probability for the type-1 Invader, while $\gamma_{i_*}^2 \omega / \gamma_{i_*}^1$ is detection probability for the type-2 Invader.

Proof. Note that $u_S^{PT}(\boldsymbol{x}, \boldsymbol{Y})$ is linear on $\boldsymbol{x}$. Then, $\boldsymbol{x}$ is the best response strategy to $\boldsymbol{Y}$ if and only if there is a ν such that

$$\alpha^1 \gamma_i^1 w_S(y_i^1) + \alpha^2 \gamma_i^2 w_S(y_i^2) \begin{cases} = \nu, & x_i > 0, \\ \leq \nu, & x_i = 0. \end{cases} \tag{47}$$

Thus, in particular, $\nu > 0$, since $\boldsymbol{x}$, $\boldsymbol{y}^1$ and $\boldsymbol{y}^2$ are probability vectors.

Note that $u_I^{PT,t}(\boldsymbol{x}, \boldsymbol{y}^t)$ is linear on $\boldsymbol{y}^t$. Thus, $\boldsymbol{y}^t$ is the best response strategy to $\boldsymbol{x}$ if and only if there is an ω^t such that

$$\gamma_i^t w_I(x_i) \begin{cases} = \omega^t, & y_i^t > 0, \\ \geq \omega^t, & y_i^t = 0, \end{cases} \tag{48}$$

where $t = 1, 2$. Thus, in particular, $\omega^1 > 0$ and $\omega^2 > 0$, since $\boldsymbol{x}$, $\boldsymbol{y}^1$ and $\boldsymbol{y}^2$ are probability vectors.

Assuming that there is an i such that $x_i = 0$ (48) implies that $\omega^t = 0$ for $t = 1, 2$. Thus, since $\boldsymbol{y}^1 = 0$ and $\boldsymbol{y}^2$ are probability vectors, (48) implies that $\boldsymbol{y}^1 \equiv \boldsymbol{y}^2 \equiv 0$. This contradiction yields that $x_i > 0$ and

$$\alpha^1 \gamma_i^1 w_S(y_i^1) + \alpha^2 \gamma_i^2 w_S(y_i^2) = \nu \text{ for } i = 1, \dots, n. \tag{49}$$

While, by (47), the assumption that there is an i such that $y_i^1 = 0$ and $y_i^2 = 0$ yields that $x_i = 0$. This contradiction implies that there is no i such that $y_i^1 = 0$ and $y_i^2 = 0$. Then, by (47) and (48) we have that

$$y_i^1 = \begin{cases} w_S^{-1}\left(\dfrac{\nu}{\alpha^1 \gamma_i^1}\right), & i \in I_{10}, \\ \alpha^1 \gamma_i^1 w_S(y_i^1) + \alpha^2 \gamma_i^2 w_S(y_i^2) = \nu, & i \in I_{11}, \\ 0, & i \in I_{01}, \end{cases} \tag{50}$$

$$y_i^2 = \begin{cases} 0, & i \in I_{10}, \\ \alpha^1 \gamma_i^1 w_S(y_i^1) + \alpha^2 \gamma_i^2 w_S(y_i^2) = \nu, & i \in I_{11}, \\ w_S^{-1}\left(\dfrac{\nu}{\alpha^2 \gamma_i^2}\right), & i \in I_{01}, \end{cases} \tag{51}$$

where
$$\begin{aligned} I_{10} &:= \{i : y_i^1 > 0, y_i^2 = 0\}, \\ I_{11} &:= \{i : y_i^1 > 0, y_i^2 > 0\}, \\ I_{01} &:= \{i : y_i^1 = 0, y_i^2 > 0\}. \end{aligned} \tag{52}$$

By (48), we have that:

(a) if $i \in I_{10}$ then

$$\gamma_i^1 w_I(x_i) = \omega^1 \tag{53}$$

and

$$\gamma_i^2 w_I(x_i) \geq \omega^2 \tag{54}$$

Dividing (54) by (53) implies

$$\gamma_i^2 / \gamma_i^1 \geq \omega^2 / \omega^1 \text{ for } i \in I_{10}. \tag{55}$$

(b) if $i \in I_{01}$ then

$$\gamma_i^1 w_I(x_i) \geq \omega^1 \tag{56}$$

and

$$\gamma_i^2 w_I(x_i) = \omega^2. \tag{57}$$

Dividing (56) by (57) implies

$$\omega^2/\omega^1 \geq \gamma_i^2/\gamma_i^1 \text{ for } i \in I_{01}. \tag{58}$$

(c) if $i \in I_{11}$ then

$$\gamma_i^1 w_I(x_i) = \omega^1 \tag{59}$$

and

$$\gamma_i^2 w_I(x_i) = \omega^2. \tag{60}$$

Dividing (60) by (59) implies

$$\omega^2/\omega^1 = \gamma_i^2/\gamma_i^1 \text{ for } i \in I_{11}. \tag{61}$$

Then, by (59), (60), (61) and assumption (27), there exists a unique i such that

$$I_{11} = \{i\}, I_{10} = \{1, \ldots, i-1\} \text{ and } I_{01} = \{i+1, \ldots, n\}.$$

Then, by (48),

$$x_j = x_j(\omega_1, \omega_2) = \begin{cases} w_I^{-1}\left(\dfrac{\omega^1}{\gamma_j^1}\right), & j \leq i-1, \\[2ex] w_I^{-1}\left(\dfrac{\omega^1}{\gamma_i^1}\right) = w_I^{-1}\left(\dfrac{\omega^2}{\gamma_i^2}\right), & j = i, \\[2ex] w_I^{-1}\left(\dfrac{\omega^2}{\gamma_i^2}\right), & j \geq i+1. \end{cases} \tag{62}$$

Let $\omega = \omega^1$. Then, by (61), $\omega^2 = \gamma_i^2 \omega/\gamma_i^1$. Thus, by (62), x has to be given by (41) with i instead of i_*. Since x is probability vector, then ω is uniquely defined by (42).

Next, we have to prove that $i = i_*$ and find ν. By (50) and (51), taking into account notations (29) and (30), and the fact that y^1 and y^2 are probability vectors, we have that i and ν are given by the following conditions:

$$\Phi_i(\nu) = \nu, \tag{63}$$
$$\underline{\Psi}_i(\nu) \leq 1 \tag{64}$$

and

$$\overline{\Psi}_i(\nu) \leq 1. \tag{65}$$

Note that

$$\Phi_i(\nu) - \nu \text{ is continuously decreasing for } \nu \in [0, \overline{\nu}_i]. \tag{66}$$

Moreover,

$$(\Phi_i(\nu) - \nu)\Big|_{\nu=0} = \alpha^1\gamma_i^1 + \alpha^2\gamma_i^2 > 0, \tag{67}$$

Thus, by (66) and (67), Eq. (63) has the root (and it is unique) in $[0, \overline{\nu}_i]$ if and only if

$$(\Phi_i(\nu) - \nu)\Big|_{\nu=\overline{\nu}_i} \leq 0. \tag{68}$$

To derive the necessary and sufficient conditions for (68) to hold, let us consider three cases separately: (i) $\nu_i^1 = \nu_i^2$, (ii) $\overline{\nu}_i = \nu_i^1 < \nu_i^2$ and (iii) $\overline{\nu}_i = \nu_i^2 < \nu_i^1$.

(i) Let $\nu_i^1 = \nu_i^2$. Then,

$$(\Phi_i(\nu) - \nu)\Big|_{\nu=\nu_i^1} = -\nu_i^1 = -\nu_i^2 < 0. \tag{69}$$

Thus, in case (i), (68) always holds. Moreover, by Proposition 1(b) and Proposition 1(d), (64) and (65) hold for each $\nu \leq \nu_i^1 = \nu_i^2$.

(ii) Let $\overline{\nu}_i = \nu_i^1 < \nu_i^2$. Then,

$$(\Phi_i(\nu) - \nu)\Big|_{\nu=\nu_i^1} = \alpha^2\gamma_i^2 w_S\left(1 - \sum_{j=i+1}^{n} w_S^{-1}\left(\frac{\nu_i^1}{\alpha^2\gamma_j^2}\right)\right) - \nu_i^1, \tag{70}$$

Thus, (68) with $\nu = \nu_i^1$ is equivalent to

$$1 \leq \sum_{j=i}^{n} w_S^{-1}\left(\frac{\nu_i^1}{\alpha^2\gamma_j^2}\right), \tag{71}$$

Moreover, taking into account notation (30) and combining (71) with the fact that (64) and (65) have to hold, we have that the following condition must hold

$$\overline{\Psi}_i(\nu_i^1) < 1 \leq \overline{\Psi}_{i-1}(\nu_i^1) \text{ with } \nu_i^1 < \nu_i^2 \tag{72}$$

By Proposition 1(d), (72) is equivalent to

$$\nu_{i-1}^2 < \nu_i^1 < \nu_i^2. \tag{73}$$

(iii) Let $\overline{\nu}_i = \nu_i^2 < \nu_i^1$. Then,

$$(\Phi_i(\nu) - \nu)\Big|_{\nu=\nu_i^2} = \alpha^1\gamma_i^1 w_S\left(1 - \sum_{j=1}^{i-1} w_S^{-1}\left(\frac{\nu_i^2}{\alpha^1\gamma_j^1}\right)\right) - \nu_i^2. \tag{74}$$

Thus, (68) with $\nu = \nu_i^2$ equivalent to

$$1 \leq \sum_{j=1}^{i} w_S^{-1}\left(\frac{\nu_i^2}{\alpha^1\gamma_j^1}\right). \tag{75}$$

Moreover, taking into account notation (30) and combining (71) with the fact that (64) and (65) must hold, we have that the following condition has to hold

$$\underline{\Psi}_i(\nu_i^2) \leq 1 < \underline{\Psi}_{i+1}(\nu_i^2) \text{ with } \nu_i^2 < \nu_i^1. \tag{76}$$

By Proposition 1(b), (76) is equivalent to

$$\nu_{i+1}^1 < \nu_i^2 < \nu_i^1 \tag{77}$$

Then, the result follows from (73), (77) and Proposition 4. ∎

Of course, the scenario with rational rivals is a boundary case for subjective behavior with $\kappa = 1$. Namely, for rational rivals the following result holds.

Corollary 2. *In the Bayesian game with a rational Scanner and rational Invader, the equilibrium strategies $(\boldsymbol{x}, \boldsymbol{Y})$ are unique and given as follows:*

$$x_i = \frac{1}{\sum_{j \leq i_*} \frac{1}{\gamma_j^1} + \sum_{j \geq i_*+1} \frac{\gamma_{i*}^2}{\gamma_{i*}^1 \gamma_j^2}} \begin{cases} \frac{1}{\gamma_i^1}, & j \leq i_*, \\ \frac{\gamma_{i*}^2}{\gamma_{i*}^1 \gamma_i^2}, & j \geq i_*+1, \end{cases} \tag{78}$$

and

$$y_i^1 = \begin{cases} \frac{\nu}{\alpha^1 \gamma_i^1}, & i \leq i_* - 1, \\ 1 - \sum_{j=1}^{i_*-1} \frac{\nu}{\alpha^1 \gamma_j^1}, & i = i_*, \\ 0, & i \geq i_*+1, \end{cases} \tag{79}$$

$$y_i^2 = \begin{cases} 0, & i \leq i_* - 1, \\ 1 - \sum_{j=i_*+1}^{n} \frac{\nu}{\alpha^2 \gamma_j^2}, & i = i_*, \\ \frac{\nu}{\alpha^2 \gamma_i^2}, & i \geq i_*+1, \end{cases} \tag{80}$$

where

$$\nu = \frac{\alpha^1 \gamma_{i*}^1 + \alpha^2 \gamma_{i*}^2}{1 + \sum_{j \leq i_*-1} \frac{\gamma_{i*}^1}{\gamma_j^1} + \sum_{j \geq i_*+1} \frac{\gamma_{i*}^2}{\gamma_j^2}} \tag{81}$$

and i_* is given by Proposition 4 with

$$\nu_i^1 = \frac{\alpha^1}{\sum_{j \leq i-1}(1/\gamma_j^1)} \text{ and } \nu_i^2 = \frac{\alpha^2}{\sum_{j \geq i+1}(1/\gamma_j^2)}. \tag{82}$$

Proof. For rational rivals $w_I(p) \equiv w_S(p) \equiv p$. Thus, by (29) and (30),

$$\underline{\Psi}_i(\nu) = \sum_{j=1}^{i-1} \frac{\nu}{\alpha^1 \gamma_j^1} \text{ and } \overline{\Psi}_i(\nu) = \sum_{j=i+1}^{n} \frac{\nu}{\alpha^2 \gamma_j^2}. \tag{83}$$

This, jointly with (32) and (33), implies (82).
By

$$x_j = x_j(\omega) = \omega \times \begin{cases} \dfrac{1}{\gamma_j^1}, & j \leq i_*, \\ \dfrac{\gamma_{i*}^2}{\gamma_{i*}^1 \gamma_j^2}, & j \geq i_* + 1. \end{cases} \tag{84}$$

Then, taking into account that x is a probability vector we obtain (78). Finally, by (85)

$$\Phi_i(\nu) = \alpha^1 \gamma_i^1 \left(1 - \sum_{j=1}^{i-1} \frac{\nu}{\alpha^1 \gamma_j^1} \right) + \alpha^2 \gamma_i^2 \left(1 - \sum_{j=i+1}^{n} \frac{\nu}{\alpha^2 \gamma_j^2} \right). \tag{85}$$

Substituting this $\Phi_i(\nu)$ into (45), and solving this equation we obtain (81), and the result follows. ∎

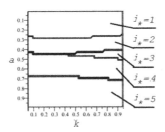

Fig. 3. Switching band i_* as function on a priori probability and probability weighting parameter κ.

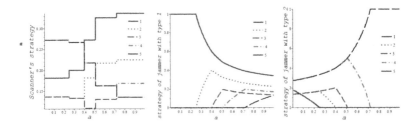

Fig. 4. The user's strategy (left), strategy of type-1 jammer (middle) and (c) strategy of type-2 jammer (right) for $\kappa = 1$.

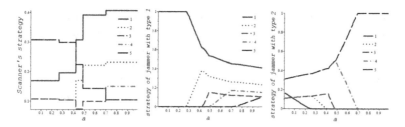

Fig. 5. The user's strategy (left), strategy of type-1 jammer (middle) and (c) strategy of type-2 jammer (right) for $\kappa = 0.75$.

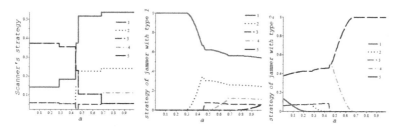

Fig. 6. The user's strategy (left), strategy of type-1 jammer (middle) and (c) strategy of type-2 jammer (right) for $\kappa = 0.5$.

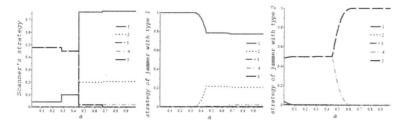

Fig. 7. The user's strategy (left), strategy of type-1 jammer (middle) and (c) strategy of type-2 jammer (right) for $\kappa = 0.25$.

Let us illustrate the obtained result by a bandwidth scanning example consisting of $n = 5$ bands, where

$$\gamma^1 = (0.2, 0.3, 0.5, 0.4, 0.5) \text{ and } \gamma^2 = (0.3, 0.4, 0.4, 0.2, 0.2).$$

Figure 3 illustrates the dependence of the switching band i_*, i.e., the band where both types of Invader sneak through with positive probability, on a priori probability $\alpha^1 = \alpha$ (thus, $\alpha^2 = 1 - \alpha$) and probability weighting parameter κ.

Figures 4, 5, 6 and 7 illustrate the dependence of the equilibrium PT strategy of the Scanner and the Invader on the a priori probability α and probability weighting parameter $\kappa \in \{0.25, 0.5, 0.75, 1\}$. The Scanner's equilibrium strategy has water-filling form (41) with water-filling equation (42). Each band is scanned with positive probability although such probabilities could tend to zero as the probability weighting parameter decreases. By (41), the Scanner's strategy x is piecewise constant with respect to a priori probability, while by (43) and (44), the Invader's strategies y^1 and y^2 are continuous with respect to a priori probability. Thus, due to such piecewise constant structure, the Scanner strategy is less sensitive to a priori probabilities compared with the Invader's strategy, except for a finite set S of a priori probabilities where the Scanner strategy has jump discontinuities. Within this finite set S, the Scanner strategy is over-sensitive to a priori probability (which is reflected by the corresponding jumps) to compensate for the lack of such sensitivity outside of the set S. Finally, the Scanner's strategy and Invader's strategies are continuous with respect to the probability weighting parameter and, in this sense, they are equally sensitive to the subjective factor reflected by the probability weighting parameter.

9   Conclusions

In this paper we have investigated the impact of subjectivity on the rival's behavior in a bandwidth scanning problem. Namely, we have formulated a prospect theoretic extension of a Bayesian game between a Scanner and an Invader where the Scanner knows only a priori probabilities about what type of intrusion (regular intensity or low intensity) occurs in bandwidth. Existence and uniqueness of the PT Bayesian equilibrium is proven. Moreover, these PT Bayesian equilibrium strategies are derived in closed form as functions of the detection probabilities. Waterfilling equations were found that allows one to derive these detection probabilities. In particular, the waterfilling equations provide a means to identify the bands where the Invader's strategies have band-sharing form, and to establish equal sensitivity of the Scanner strategy and the Invader strategy to the subjective factor reflected by the probability weighting parameter. It is worth remarking that, the rival strategies can have different sensitivity with respect to a priori probabilities about the intrusion type. Namely, the Scanner strategy is piecewise constant while the Invader's strategy is continuous with respect to such a priori probability. Thus, the Scanner strategy combines non-sensitive behavior (when it is constant) with over sensitive behavior (when it has jumps). The goal of our future research is to generalize the obtained result for more general cases regarding the intrusion types.

References

1. Altman, E., Avrachenkov, K., Garnaev, A.: Generalized α-fair resource allocation in wireless networks. In: 47th IEEE Conference on Decision and Control (CDC 2008), Cancun, Mexico, pp. 2414–2419 (2009)
2. Anindya, I.C., Kantarcioglu, M.: Adversarial anomaly detection using centroid-based clustering. In: IEEE International Conference on Information Reuse and Integration (IRI), pp. 1–8 (2018)
3. Baston, V.J., Garnaev, A.Y.: A search game with a protector. Naval Res. Logistics **47**, 85–96 (2000)

4. Comaniciu, C., Mandayam, N.B., Poor, H.V.: AWireless Networks Multiuser Detection in Cross-Layer Design. Springer, New York (2005)
5. Dambreville, F., Le Cadre, J.P.: Detection of a markovian target with optimization of the search efforts under generalized linear constraints. Naval Res. Logistics **49**, 117–142 (2002)
6. Digham, F.F., Alouini, M.S., Simon, M.K.: On the energy detection of unknown signals over fading channels. IEEE Trans. Commun. **55**, 21–24 (2007)
7. Garnaev, A.: A remark on a helicopter and submarine game. Naval Res. Logistics **40**, 745–753 (1993)
8. Garnaev, A., Garnaeva, G., Goutal, P.: On the infiltration game. Int. J. Game Theory **26**, 215–221 (1997)
9. Garnaev, A., Trappe, W.: One-time spectrum coexistence in dynamic spectrum access when the secondary user may be malicious. IEEE Trans. Inf. Forensics Secur. **10**, 1064–1075 (2015)
10. Garnaev, A., Trappe, W.: A bandwidth monitoring strategy under uncertainty of the adversary's activity. IEEE Trans. Inf. Forensics Secur. **11**, 837–849 (2016)
11. Garnaev, A., Trappe, W., Kung, C.-T.: Optimizing scanning strategies: selecting scanning bandwidth in adversarial RF environments. In: 8th International Conference on Cognitive Radio Oriented Wireless Networks (Crowncom), pp. 148–153 (2013)
12. Guan, S., Wang, J., Jiang, C., Tong, J., Ren, Y.: Intrusion detection for wireless sensor networks: a multi-criteria game approach. In: IEEE Wireless Communications and Networking Conference (WCNC), pp. 1–6 (2018)
13. Han, Z., Niyato, D., Saad, W., Basar, T., Hjrungnes, A.: Game Theory in Wireless and Communication Networks: Theory, Models, and Applications. Cambridge University Press, New York (2012)
14. Hohzaki, R.: An inspection game with multiple inspectees. Eur. J. Oper. Res. **178**, 894–906 (2007)
15. Hohzaki, R., Iida, K.: A search game with reward criterion. J. Oper. Res. Soc. Japan **41**, 629–642 (1998)
16. Jotshi, A., Batta, R.: Search for an immobile entity on a network. Eur. J. Oper. Res. **191**, 347–359 (2008)
17. Kahneman, D., Tversky, A.: Prospect theory: an analysis of decision under risk. Econometrica **47**, 263–291 (1979)
18. Kahneman, D., Tversky, A.: Advances in prospect theory: cumulative representation of uncertainty. J. Risk Uncertainty **5**, 297–323 (1992)
19. Liu, S., Chen, Y., Trappe, W., Greenstein, L.J.: ALDO: an anomaly detection framework for dynamic spectrum access networks. In: IEEE International Conference on Computer (INFOCOM), pp. 675–683 (2009)
20. Poongothai, T., Jayarajan, K.: A noncooperative game approach for intrusion detection in mobile adhoc networks. In: International Conference on Computing, Communication and Networking, pp. 1–4 (2008)
21. Poor, H.V.: An Introduction to Signal Detection and Estimation. Springer, New York (1994). https://doi.org/10.1007/978-1-4757-2341-0
22. Prelec, D.: The probability weighting function. Econometrica **90**, 497–528 (1998)
23. Reddy, Y.B.: A game theory approach to detect malicious nodes in wireless sensor networks. In: Third International Conference on Sensor Technologies and Applications, pp. 462–468 (2009)
24. Saad, W., Sanjab, A., Wang, Y., Kamhoua, C.A., Kwiat, K.A.: Hardware trojan detection game: a prospect-theoretic approach. IEEE Trans. Veh. Technol. **66**, 7697–7710 (2017)

25. Sakaguchi, M.: Two-sided search games. J. Oper. Res. Soc. Japan **16**, 207–225 (1973)

26. Sanjab, A., Saad, W., Basar, T.: Prospect theory for enhanced cyber-physical security of drone delivery systems: a network interdiction game. In: IEEE International Conference on Communications (ICC), Paris, France (2017)

27. Sauder, D.W., Geraniotis, E.: Signal detection games with power constraints. IEEE Trans. Inf. Theory **40**, 795–807 (1994)

28. Shen, S.: A game-theoretic approach for optimizing intrusion detection strategy in WSNs. In: 2nd International Conference on Artificial Intelligence, Management Science and Electronic Commerce (AIMSEC) (2011)

29. Vamvoudakis, K.G., Hespanha, J.P., Sinopoli, B., Mo, Y.: Adversarial detection as a zero-sum game. In: IEEE 51st IEEE Conference on Decision and Control (CDC), pp. 7133–7138 (2012)

30. Wang, X., Feng, R., Wu, Y., Che, S., Ren, Y.: A game theoretic malicious nodes detection model in MANETs. In: IEEE 9th International Conference on Mobile Ad-Hoc and Sensor Systems (MASS), pp. 1–6 (2012)

31. Xiao, L., Liu, J., Li, Q., Mandayam, N.B., Poor, H.V.: User-centric view of jamming games in cognitive radio networks. IEEE Trans. Inf. Forensics Secur. **10**, 2578–2590 (2015)

32. Xiao, L., Liu, J., Li, Y., Mandayam, N.B., Poor, H.V.: Prospect theoretic analysis of anti-jamming communications in cognitive radio networks. In: IEEE Global Communications Conference, pp. 746–751 (2014)

33. Xu, D., Xiao, L., Mandayam, N.B., Poor, H.V.: Cumulative prospect theoretic study of a cloud storage defense game against advanced persistent threats. In: IEEE Conference on Computer Communications Workshops (INFOCOM WKSHPS), pp. 541–546 (2017)

Games and Random Search

Artur Popławski[1,2]([⊠])

[1] NOKIA Kraków Technology Center, Kraków, Poland
artur.poplawski@nokia.com
[2] Department of Telecommunications, AGH University of Science and Technology,
Kraków, Poland

Abstract. We use dynamics of measures, i.e. iteration of the operators from measurable space to space of probabilistic measures on this space, to model and prove properties of random search algorithms. Specifically using this technique in the context of Game Theory we show that stochastic better response dynamics, where players in the potential game perform their moves independently choosing the random strategy improving their outcome, converges in stochastic sense to playing strategies near equilibrium.

Keywords: Game Theory · Random search · Measure dynamics

1 Introduction

Main aim of the paper is to study an effective method of mathematical analysis of random searching as dynamics in the space of probabilistic measures. This method allows to relatively easily prove the convergence of the random search algorithms, also those realized in multi-agent scenarios. Although work was conducted with applications to wireless networks in mind, discussion is abstract making its outcomes applicable in the broad context. Specifically, we will show correspondence between "classical" random search and random search, as may be realized in Game Theoretic problem of finding equilibria. Formalism used to capture stochasticity here is inspired by [1] and applications of measure dynamics for proving convergence of the random search algorithms in [4] and [6] (where it is applied to much more sophisticated algorithms than considered in this work).

The paper has a following structure. First, we introduce the method of measure dynamics by presenting some basic facts and a simple example of random search in optimization. Results in this section largely overlap with results in [5], but the formal approach is different. In the next section, we pose the problem of finding equilibrium of a game as an optimization problem, introducing some auxiliary function associated with a game and proving some of its basic properties. Main result, i.e. convergence of the stochastic best response dynamics in potential games to equilibria as anticipated in [3], is proven in the next section. In the last section, we comment on the main result, discuss assumptions under which it is proven and outline further development and applications. The appendix

K. Avrachenkov et al. (Eds.): GameNets 2019, LNICST 277, pp. 29–47, 2019.
https://doi.org/10.1007/978-3-030-16989-3_3

contains some important facts and proofs required for the completeness of the proof of the main theorem but are not essential from the point of view of its exposition. We will use following notation and conventions in the text:

- $U_\epsilon = f^{-1}((\epsilon, \infty))$ for $f : S \to \mathbb{R}$, where S and f is known from context. In text care will be taken to clearly specify this context.
- For $s \in \prod_{j \in P} S_j$ and $x \in S_i$ we have $(x, s_{-i}) = y \in \prod_{j \in P} S_j$, such that $s_j = y_j$ for $j \neq i$ and $x = y_i$.
- For $s \in \prod_{j \in P} S_j = S$ we will denote by $\iota_{s,i} : S_i \to S$ embedding defined as $\iota_{s,i}(x) = (x, s_{-i})$.
- For a measurable space (X, σ) by $\mathcal{P}(X)$ we will denote the set of all probabilistic measures on X.

2   Stochastic Search

Random search is the well established, generic method of solving optimization problem or finding an object satisfying some special criteria. The latter is to some extent a special case of the former, since one can think of finding objects as of finding maxima of a characteristic function. This random search method is also known as stochastic optimization or stochastic search. It is an important component of methods known in Operational Research as Metaheuristics [2].

There are numerous variants of the method. In this section, we will only consider one, which is the easiest for formal analysis. Other will be only mentioned. For more sophisticated examples one can refer e.g. to [4] or [6] and to literature referred by these works.

Let (X, σ) be a measurable space (σ is a σ-algebra on X). Let f be a function $f : X \to \mathbb{R}$ Let μ_0 be some probabilistic measure on (X, σ) being the "reference" measure in further discussion. Let $\mathcal{P}(X)$ be the set of probabilistic measures on (X, σ) General problem we are trying to solve is to find $x^* \in X$ such that certain maximality condition is satisfied. Exact form of this condition depends on the quality of the solution we want to achieve. Possible forms of such a condition are:

$f(x^*) = \max_{x \in X} f(x)$ (if x^* exists, e.g. X is finite or X is compact and f is continuous)

$f(x^*) > \sup_{x \in X} f(x) - \epsilon$ for some $\epsilon > 0$

$\mu_0(\{x : f(x) > f(x^*)\}) < \epsilon$ for some $\epsilon > 0$

Last condition, which amounts to finding the solution difficult to beat, is not vastly discussed in the literature, however it seems to be potentially important in certain applications like games, risk management, security etc.

Essentials of the method lie in finding stochastic (Markov) operator

$$A : X \to \mathcal{P}(X)$$

such that condition similar to:

input : terminating condition: $C : \mathbb{R} \times \mathbb{N} \to \{True, False\}$
input : objective function: $f : X \to \mathbb{R}$
input : Initial point: $x_0 \in X$
output: $x^* \in X$
initialize variables: $x = x_0$;
$n = 0$;
while $C(f(x), n) = False$ **do**
$\quad\mid\quad x = sample(A(x))$;
$\quad\mid\quad n = n + 1$;
end
return x

Algorithm 1. Basic random search

$$\mathbb{E}_{A(x)}[f] = \int_X f dA(x) > f(x)$$

is satisfied sufficiently often. For reference we will call this condition "martingal" condition. In Operational Research setting, one can think of A as about formal representation of heuristics. Further assumptions clarifying meaning of vague term "often" and introducing conditions stronger than simple inequality depend on the particular method and the context.

Assuming that the operator A is given, and for $x \in X$ there is a method of sampling the space X according to probability measure $A(x)$ one can give computational procedure as Algorithm 1. In more precise formulation this means, that given $A(x)$ one has also $S(x)$ which is a random variable with values in the space X and distribution $A(x)$. In practice, it is most often the case that sampling method for $A(x)$ is derived from the given sampling method for measure μ_0 - see e.g. [5][1].

It is worth to note, that non-stochastic search can be treated as a special case of the stochastic search, where $A(x) = \delta_{D(x)}$ where $D : X \to X$ is a deterministic operator given by the search method and δ_x is a proper Dirac's measure, (i.e. $\int_X g d\delta_x = g(x)$ for each $g : X \to \mathbb{R}$).[2]

---

[1] Opposite direction, where one has simple method of sampling for $A(x)$ but sampling X is difficult, is also used in practice. When μ_0 is stationary distribution of Markov process given by A, famous Metropolis algorithm is such an example.

[2] To some extent one can look at the stochastic search method as having significant advantage over the deterministic one, because deterministic method relies on some internal structure of the X. If function we are maximizing is incompatible with this internal structure, deterministic search may be completely lost. A method of sampling assumed is in this case a way to escape from this pitfall. On the other hand, ability to sample requires some source of randomness - using of pseudo-random variables for sampling makes the method de facto deterministic. Intuition when such a method cannot work efficiently comes from Algorithmic Information Theory. In purely computable setting there is no randomness. This is an interesting philosophical aspect of random search.

Example 1 (Random search). For $v \in \mathbb{R}$ let's denote $U_v = \{x \in X : f(x) > v\}$ and define operator A (remembering, that $A(x)$ is a measure):

$$A(x)(B) = \begin{cases} \mu_0(B \cap U_{f(x)}) + \mu_0(X - U_{f(x)}) & x \in B \\ \mu_0(B \cap U_{f(x)}) & otherwise \end{cases}$$

Please note, that this operator corresponds to $A(x)$ sampling procedure that uses sampling for the μ_0. We can, with some small abuse of the notation, express it as:

$$sample(A(x)) = \begin{cases} y & y = sample(\mu_0) \text{ and } f(y) > f(x) \\ x & y = sample(\mu_0) \text{ and } f(y) \leq f(x) \end{cases}$$

Less formally we can say, that at ech step we are drawing an element according to distribution μ_0 and return this element if the value of f on this element is greater than $f(x)$ or return x otherwise.

We can check in this case:

$$\mathbb{E}_{A(x)}[f] = \int_X f dA(x) = \int_{X - U_{f(x)}} f dA(x) + \int_{U_{f(x)}} f dA(x) =$$

$$f(x)\mu_0(X - U_{f(x)}) + \int_{U_{f(x)}} f dA(x) \geq f(x)\mu_0(X - U_{f(x)}) + f(x)\mu_0(U_{f(x)}) = f(x)$$

If additionally $\mu_0(U_{f(x)}) > 0$ aforementioned weak inequality becomes strong. So to have "martingal" condition satisfied one has to assume something about f (to satisfy this "sufficiently often" part). E.g. assumption that $\mu_0(\{x \in X : \mu_0(U_{f(x)}) > 0\}) = 1$ may be a good in this case.

As the extension of Example 1 we can assume stronger conditions on X postulating additional structure e.g. that X is topological or metric space and require from f to be compatible with this structure: e.g. to be continuous.

Example 2 (Simple local search). Let's assume, that X is sufficiently regular topological space (e.g. metrical), where τ denotes topology and let σ be a Borrel's σ-algebra (i.e. the smallest σ-algebra containing all open sets). Let $N : X \to \tau$ be such, that $x \in N(x)$. We can define operator $A : X \to \mathcal{P}(X)$:

$$A(x)(B) = \begin{cases} \frac{1}{\mu_0(N(x))}\mu_0(B \cap U_{f(x)} \cap N(x)) + \mu_0((X - U_{f(x)}) \cap B \cap N(x)) & x \in B \\ \frac{1}{\mu_0(N(x))}\mu_0(B \cap U_{f(x)} \cap N(x)) & otherwise \end{cases}$$

This operator corresponds to "local search" where sampling is performed over some neighborhood $N(x)$ of x. Additional assumptions about $N(x)$ should be made to allow Algorithm 1 to escape from local maxima. Particularly bad choice of X and N may result in not all the space being reachable (so, in the language of the Markov chains, process is not irreducible).

input : terminating condition: $C : \mathbb{R} \times \mathbb{N} \to \{True, False\}$
input : objective function: $f : X \to \mathbb{R}$
input : Initial point: $x_0 \in X$
output: $x^* \in X$
initialize variables: $x = x_0$;
$n = 0$;
$s = s_0$ **while** $C(f(x), n) = False$ **do**
 | $x = sample(\pi_1(A(x, s)))$;
 | $s = \pi_2(A(x, s))$;
 | $n = n + 1$;
end
return x

Algorithm 2. State random search

In both examples, the choice of the operators A prevented sampling process from decreasing value of the $f(x)$ along the trajectory constituted by selection of subsequent xs in the Algorithm 1. This may not necessarily be the case in random searching. E.g it is not the case in the famous simulated annealing algorithm, where there are possible some moves (choices of the points) decreasing optimized function. Their probability however as a function of number of the step in which the choice is made is decreasing. This means also, that in this kind of algorithm we do not have a fixed A operator, but we use different operators in subsequent steps.

To properly explore this extended setting, one, following e.g. [2], may extend definition by introducing state space. We take (X, σ) as previously, but introduce space S which we can think of as about some "memory" or "learning ability". We take operator $A : X \times S \to \mathcal{P}(X) \times S$. We can now state Algorithm 2.

In the Algorithm 2 π_i means projection on the i-th component in the Cartesian product. Let's observe, that previous, stateless operators A can be treated as special cases of these new A when we consider dummy, single element, state space $\{*\}$ and define $A(x, *) = (A(x), *)$ We can also fit into this model case, where A changes as a function of step as in the simulated annealing. In this case operator is of the form $A(x, n) = (A_n(x), n + 1)$.

Now, let's analyze issues of convergence of the algorithms. Let's define measures μ_i for $i > 0$ inductively by defining for $B \in \sigma$ $\mu_{n+1}(B) = \int_X A_n(x)(B) d\mu_n$

Under some technical conditions, that we will not discuss for this case, by Fubini's theorem μ_i is well defined and is indeed a measure. We have:

Theorem 1. *If for each i, $A_i : X \to \mathcal{P}(X)$ is such, that $\mathbb{E}_{A_i(x)}[f] > f(x)$ then for each i $\mathbb{E}_{\mu_{i+1}}[f] \geq \mathbb{E}_{\mu_i}[f]$.*

Proof. Let $f_k : X \to \mathbb{R}$ be a function such that $f_k|_{B_k} \equiv c_k$ and $f_k|_{X - B_k} \equiv 0$ for certain $B_k \in \sigma$, $c_k \in \mathbb{R}$. For any ϵ there is countable K and family $\{B_k\}_{k \in K}$ such that for each $k \in K$ $B_k \in \sigma$ and sets are pairwise mutually exclusive, such that step function $F_\epsilon = \sum_{k \in K} f_k \leq f$ and $\int_X f d\mu_i - \int_X F_\epsilon d\mu_i \leq \epsilon$ and $\int_X f d\mu_{i+1} - \int_X F_\epsilon d\mu_{i+1} \leq \epsilon$ Now, taking some fixed ϵ we have:

$$\int_X f d\mu_{i+1} > \int_X F_\epsilon d\mu_{i+1} = \int_X (\sum_{k \in K} f_k) d\mu_{i+1} = \sum_{k \in K} \int_X f_k d\mu_{i+1} =$$

$$\sum_{k \in K} c_k \mu_{i+1}(B_k) = \sum_{k \in K} c_k \int_X A_i(x)(B_k) d\mu_i =$$

$$\int_X (\sum_{k \in K} c_k A_i(x)(B_k)) d\mu_i = \int_X (\int_X (\sum_{k \in K} f_k) dA_i(x)) d\mu_i$$

Taking subsequent approximation f by step functions $\sum_{k \in K} f_k$ we have the last expression converging to

$$\int_X \int_X f dA_i(x) d\mu_i = \int_X \mathbb{E}_{A_i(x)}[f] d\mu_i \geq \int_X f d\mu_i = \mathbb{E}_{\mu_i}[f]$$

We also have following form of convergence of measures to maximal solution:

Theorem 2. *Let's fix $\gamma > 0$. If for each i and each x $A_i(x)(U_{max-\epsilon}) > \gamma$ and $A_i(x)(X - U_{f(x)}) = A(x)(x)$ then $\mu_i(U_{max-\epsilon}) \underset{i \to \infty}{\longrightarrow} 1$.*

Proof.

$$\mu_{i+1}(U_{max-\epsilon}) = \int_X A_i(x)(U_{max-\epsilon}) d\mu_i =$$

$$\int_{U_{max-\epsilon}} A_i(x)(U_{max-\epsilon}) d\mu_i + \int_{X - U_{max-\epsilon}} A_i(x)(U_{max-\epsilon}) d\mu_i \geq$$

$$\int_{U_{max-\epsilon}} 1 d\mu_i + \int_{X - U_{max-\epsilon}} \gamma d\mu_i = \mu_i(U_{max-\epsilon}) + \gamma(1 - \mu_i(U_{max-\epsilon})) =$$

$$(1 - \gamma)\mu_i(U_{max-\epsilon}) + \gamma$$

Now we can inductively obtain:

$$\mu_i(U_{max-\epsilon}) \geq (1-\gamma)^i \mu_0(U_{max-\epsilon}) + \gamma(\sum_{k=0}^{i-1}(1-\gamma)^k) = 1 + (1-\gamma)^i(\mu_0(U_{max-\epsilon}) - 1)$$

Thus $\mu_i(U_{max-\epsilon}) \underset{i \to \infty}{\longrightarrow} 1$

3   Finding Equilibrium as Optimization Problem

Let's consider, as in the Game Theory, a game in extensive form:

Definition 1 (Game). *A game* Γ *in extensive form is triple* $(P, \{S_i\}_{i \in P}, \{u_i\}_{i \in P})$ *where* P *is a (finite) set of players, for each* $i \in P$, S_i *is a set of strategies (or actions) available to player* i *and*

$$u_i : \prod_{i \in P} S_i \to \mathbb{R}$$

is a function called "payoff" of player i. *For convenience we will denote* $S = \prod_{i \in P} S_i$.

Important notion in the Game Theory is the Nash Equilibrium (NE). A point $s \in S$ is NE when no player wants to deviate from it individually. Formally

Definition 2 (Nash Equilibrium). *Having game* Γ, *under the notation of Definition 1, an element* $s \in S$ *is Nash Equilibrium if for each* $i \in P$ *and for each* $s_i' \in S_i$ $u_i(s_i', s_{-i}) \leq u_i(s)$.

As it is vastly discussed in the Game Theory literature, NE models situation which will be stable in the game assuming that players are perfectly rational, which means they want to secure highest possible payoff for themselves. We will denote:

$$N_\Gamma = \{x \in S : x \text{ is NE for } \Gamma\}$$

For game Γ let's define function $\chi_\Gamma \to \mathbb{R} \cup \{-\infty\}$:

$$\chi_\Gamma(s) = \min_{i \in P}(\min_{x_i \in S_i} (u_i(s) - u_i(x_i, s_{-i}))))$$

In general χ_Γ may be not defined for all points (or rather take infinite values as some points). In practice however we often assume more about structure of the strategies and payoff function, which makes χ_Γ behaving more regularly and having its range in the reals.

Theorem 3. *Let game* $\Gamma = (P, \{S_i\}_{i \in P}, \{u_i\}_{i \in P})$ *be such that for each* $i \in P$ (S_i, d_i) *is compact metric space. Let for each* i u_i *be continuous in topology on* S *given by the metric equivalent to product metric on* S *i.e. metric* d *such that* $d(s, s') = \sum_{i \in P} d_i(s_i, s_i')$. *Under these assumptions* $\chi_\Gamma(S) \subset \mathbb{R}$ *and* χ_Γ *is continuous.*

Proof. For each $i \in P$ and $s \in S$ function $\phi_{i,s}(x) = u_i(x, s_i) - u_i(s)$ is continuous, so because S_i is compact it attains mininimum. Thus $\chi_\Gamma(s) = \min_{i \in P} \min \phi_{i,s} \in \mathbb{R}$.

Now, let's take $s \in S$ and $\epsilon > 0$. Since S is compact each u_i and ϕ_i is uniformly continuous. There is δ_i such that $d(x, y) < \delta_i$ implies $|u_i(x) - u_i(y)| < \epsilon$ Now if $d(x, y) < \delta_i$ we have for each t $|\phi_{i,x}(t) - \phi_{i,y}(t)| < \epsilon$ (since $d((t, x_{-i}), (t, y_i) \leq$

$d(x,y))$. This implies that $|\min\phi_{i,x} - \min\phi_{i,y}| < \epsilon$ So, taking $\delta < \min\limits_{i\ in\ P}\delta_i$ we have, that for each s' such that $d(s,s') < \delta$ $|\chi_\Gamma(s) - \chi_\Gamma(s')| = |\min\limits_{i\in P}\min\phi_{i,S} - \min\limits_{i\in P}\min\phi_{i,S'}| < \epsilon$

Similarly we have following:

Theorem 4. *Let game* $\Gamma = (P, \{S_i\}_{i\in P}, \{u_i\}_{i\in P})$ *be such that for each* $i \in P$ (S_i, d_i) *is compact metric space. Let for each* i u_i *be Lipschitz with respect to metric equivalent to product metric on* S. *Under these assumptions* $\chi_\Gamma(S) \subset \mathbb{R}$ *and* χ_Γ *is Lipschitz.*

Proof. For each $i \in P$ let L_i be a Lipschitz constant for the u_i. We will give proof by contradiction. Let's assume, that χ_Γ is not Lipschitz. So for each $M > 0$ there exists $s \in S$ and $s' \in S$ such that:

$$|\chi_\Gamma(s) - \chi_\Gamma(s')| > Md(s,s') = M(\sum_{i\in P} d_i(s_i, s_i'))$$

Let's take $M > 2 \times \max\limits_{i\in P} L_i$ and s and s' such this inequality is fulfilled. Without loss of the generality we will assume, that $\chi_\Gamma(s) > \chi_\Gamma(s')$. From inequality above we have:

$$\chi_\Gamma(s') < \chi_\Gamma(s) - Md(s,s')$$

Since each u_i is continuous we can find:

$$j \in P, y \in S_j \text{ such that } \chi_\Gamma(s') = u_j(s') - u_j(y, s_{-j})$$

Now, we have:

$$u_j(s) - u_j(y, s_{-j}) = u_j(s') + (u_j(s) - u_j(s')) - u_j(y, s'_{-j}) + (u_j(y, s_{-j}) - u_j(y, s'_{-j}))$$

so:

$$u_j(s) - u_j(y, s_{-j}) \le \chi_\Gamma(s') + |u_j(s) - u_j(s')| + |u_j(y, s_{-j}) - u_j(y, s'_{-j})|$$
$$\le \chi_\Gamma(s') + 2L_j d(s', s)$$

So, from definition of χ_Γ and assumptions about s, s' and M one has:

$$\chi_\Gamma(s) \le u_j(s) - u_j(y, s_{-j}) \le \chi_\Gamma(s') + 2L_j d(s', s) <$$
$$\chi_\Gamma(s) - (M - 2L_j)d(s', s) < \chi_\Gamma(s)$$

what is a contradiction.

Obviously, assumption about the metric d being the exact product metric in previous theorems is purely technical and simplifies proofs. In fact, any metric that is equivalent to product metric will work, but inequalities to consider will be somewhat more complicated.

We have a following observation:

Theorem 5. *Let Γ be a game. $s \in \prod_{p \in P} S_i$ is NE of Γ if and only if s is maximum of χ_Γ and $\chi_\Gamma(s) = 0$.*

Proof. Let s be a NE. Let's take $i \in P$. By the Definition 2 for each $x_i \in S_i$ we have

$$u_i(x_i, s_{-i}) \le u_i(s).$$

This implies

$$\min_{x_i \in S_i} (u_i(s) - u_i(x_i, s_{-i})) \ge 0$$

so in fact

$$\min_{x_i \in S_i} (u_i(s) - u_i(x_i, s_{-i})) = 0$$

because equality is reached at $x_i = s_i$. Since we can follow this for each $i \in P$ we have for s

$$\min_{i \in P} (\min_{x_i \in S_i} (u_i(s) - u_i(x_i, s_{-i}))) = 0$$

For s which is not NE, for at least one i we have $u_i(x_i, s_{-i}) \le u_i(s)$ so $\min_{x_i \in S_i} (u_i(s) - u_i(x_i, s_{-i}))) < 0$ and in consequence $\chi_\Gamma(s) < 0$. Thus if s is NE, s is maximum of χ_Γ. Proof in opposite direction basically goes along the same lines and will be omitted.

Let's observe, that although under assumptions about game Γ as in the Theorem 3 the χ_Γ always has maximum. Value of this maximum may be different than 0. This corresponds to situation when game has no Nash Equilibrium.

Now, let's define, often more practical then NE, notion of ϵ-Nash Equilibrium:

Definition 3 (ϵ-Nash Equilibrium). *Given $\epsilon > 0$ and game Γ as in Definition 1, point $s \in S$ is ϵ-Nash Equilibrium (ϵ-NE in short) if for each $i \in P$ and each $x \in S_i$ we have $u_i(x, s_{-i}) \le u_i(s) + \epsilon$*

We may denote set of ϵ-NE of the game: $NE_{\Gamma,\epsilon} = \{s : s \text{ is } \epsilon\text{-NE}\}$ We have simple theorem:

Theorem 6. *$NE_{\Gamma,\epsilon}$ has following properties:*

1. *$NE_{\Gamma,\epsilon} = \chi_\Gamma^{-1}((-\epsilon, \infty]) = \chi_\Gamma^{-1}((-\epsilon, 0])$*
2. *$NE_\Gamma \subset NE_{\Gamma,\epsilon'} \subset NE_{\Gamma,\epsilon}$ for each ϵ, ϵ' such that $0 < \epsilon' \le \epsilon$*
3. *under assumption of Theorem 3 $NE_{\Gamma,\epsilon}$ is closed and if $\epsilon > \epsilon' > 0$ and $NE_{\Gamma,\epsilon'} \ne \emptyset$ then $int(NE_{\Gamma,\epsilon}) \ne \emptyset$ where int denotes operation of taking interior of the set.*

Proof. First two items are direct consequence of definition. Last item is consequence of first two and fact, that χ_Γ is continuous.

4   Better Response Dynamics in Potential Game

It is well known that following class of games is important from the point of view of applications (see e.g. [3]):

Definition 4 (Potential Game). *Let* Γ *be a game as in Definition 1. We call it potential game if there exists function* $V :\rightarrow \mathbb{R}$ *called potential, such that for each* $i \in P$, *for each* $x \in S_i$ *and for each* $s \in S$ *we have:*

$$u_i(x, s_{-i}) - u_i(s) = V(x, s_{-i}) - V(s)$$

We call Γ *ordered potential game if there exists* $V : S \rightarrow \mathbb{R}$, *called potential, such that for each* $i in P$ *for each* $x \in S_i$ *and for each* $s \in S$ *we have:*

$$u_i(x, s_{-i}) - u_i(s) > 0 \iff V(x, s_{-i}) - V(s) > 0$$

We have following theorem that we will leave without a proof (and we will not use it further in the paper).

Theorem 7. *If game* Γ *is like in the assumption of Theorem 3 and is a potential game, then potential* V *is continuous.*

Main aim of this section is to prove that certain way of playing (which we call stochastic better response) multiple times the same potential game leads to playing almost surely strategy that is ϵ-NE

We will assume that we have a game Γ that is potential, compact, payoffs are continuous and that potential of the game is also continuous. By stochastic better response dynamics we will call a process with discrete time T, such that at each moment in time one of the player performs a move (so unilaterally chooses strategy) in such a way that this strategy gives the same or better outcome than the strategy chosen recently assuming that adversaries play their recent strategies. We assume that a strategy is random - so "chooses" means chooses randomly from the set of strategies according to some probability measure. We assume that which player is making move at time t is random variable, that for different times variable are independent and probability of choosing any player is non-zero[3].

Let's put it in the following, more formal, framework. Let P be finite. Let (S_i, d_i) be a compact metric space for each $i \in P$. Let σ_i be a Borel σ-algebra on S_i. Let $S = \prod_{i \in P} S_i$. Obviously, there is a natural product metrics on S and σ-algebra of Borel sets for S is $\sigma = \bigotimes_{i \in P} \sigma_i$ (product σ-algebra). Let each S_i be

---

[3] To some extent it can model situation where players are continuously playing their strategies and making decisions about change of the strategy in the moments that are distributed according to a Poisson process. Probability of choice of the same moments for some players is then 0, so we can concentrate only on the order of changes.

equipped with measure $\mu_0^i \in \mathcal{P}(S_i)$ and we take measure $\bigotimes_{i \in P} \mu_0^1 = \mu_0 \in \mathcal{P}(S)$. For $s \in S$ and $i \in P$ let's define embedding function $\iota_{s,i} : S_i \to S$ by

$$\iota_{s,i}(x) = (x, s_{-i})$$

This definition induces operators of type $\tilde{\mathcal{V}}_i : S \to \mathcal{P}(S)$. For Borel sets B we define:

$$\tilde{\mathcal{V}}_i(s)(B) = \mu_0^i(\iota_{s,i}^{-1}(B))$$

Let's note, that, obviously, $\mathcal{V}_i(s)$ is no absolutely continuous relative to μ_0. Let $\alpha \in \prod_{i \in P}(0,1)$ and such that $\sum_{i \in P} \alpha_i = 1$. Note, that for each i $\alpha_i > 0$. We define $\tilde{A} : S \to \mathcal{P}(S)$ as

$$\tilde{A}(s) = \sum_{i \in P} \alpha_i \tilde{\mathcal{V}}_i(s)$$

By analogy to examples discussed in the Sect. 2, this operator models situation when at some position s one of the players (chosen according to distribution α) choses unilaterally strategy, randomly according to measure μ_0^i. By modeling here it is understood, that random choice, is lifted to the space of measures.

Now we will introduce operator A that models stochastic better response dynamics, so procedure where at each step the single random player chooses better strategy randomly. Choice for each player is performed according to the distribution which corresponds to the measure μ_0^i restricted to the set of better strategies and normalized.

For $i \in P$ and $s \in S$ let's denote:

$$U_{i,s} = \iota_{i,s}^{-1}([u_i(s), \infty)) = \{x \in S_i : u_i(x, s_{-i}) \geq u_i(s)\}$$

We take $A : S \to \mathcal{P}(S)$ defined as

$$A(s) = \sum_{i \in P} \alpha_i \mathcal{V}_i(s)$$

where

$$\mathcal{V}_i(s)(B) = \begin{cases} \frac{\tilde{\mathcal{V}}_i(s)(B \cap U_{i,s})}{\tilde{\mathcal{V}}_i(s)(U_{i,s})} & \tilde{\mathcal{V}}_i(s)(U_{i,s}) > 0 \\ 1 & \tilde{\mathcal{V}}_i(s)(U_{i,s}) = 0, s \in B \\ 0 & otherwise \end{cases}$$

Let us, as in the Sect. 2, introduce sequence of measures starting of μ_0 inductively:

$$\mu_{n+1}(B) = \int_X A_n(x)(B) d\mu_n$$

for B being the Borel set. μ_n is distribution of n-th points on the trajectory of stochastic better response dynamics, when the initial choice of starting point was made according to distribution μ_0 and subsequent choices are governed by distributions given by A.

Now we will formulate theorem, being an variant of theorem postulated in [3].

Theorem 8. *Let* Γ *be a compact ordinal potential game with the Lipschitz potential* V *and let each* S_i *is endowed with* σ-*algebra of Borel sets and probabilistic measure* μ_0^i *satisfying additional assumption* (*): *for each* $\delta > 0$, $\inf\limits_{x \in S_i} \mu_0^i(B_i(x, \delta)) > 0$.

Let $\epsilon > 0$. *We further assume that* S *is endowed with product metrics, product* σ-*algebra and product measure. Let's consider better response dynamics described by the operator* A *above. We have:* $\mu_n \xrightarrow[n \to \infty]{} 1$. $n \to 1$

Proof. For the simplification of the argument we will assume that $\min\limits_{x \in S} V(x) = 0$. We do not loose generality by making this assumption since it is not value of potential but difference of the potential in different points that matters in definition of potential game (by adding constant to potential one still obtains valid potential for the game)

Let's take $\epsilon' = \frac{\epsilon}{2}$. We define:

$$f(s) = A(s)(U_{V(s)+\epsilon'})$$

We claim, that:

1. f is measurable
2. there is $\delta > 0$ such that $f(s) > \delta$ for all s in compact set $\chi_\Gamma^{-1}((-\infty, \epsilon])$.

We will postpone technical proof of first property to Appendix. Last property follows from the fact that, as we have already observed if the $\chi_\Gamma(s) < \epsilon$ there is i and $x \in S_i$ such that $V(x, s_{-i}) - V(s) = u_i(x, s_{-i}) - u_i(s) > \epsilon$. So $V(x, s_{-i}) > V(s) + \epsilon$. From Lipschitzity of V, for $y \in B((x, s_{-i}), \frac{\epsilon'}{L})$ we have

$$V(y) > V(x, s_{-i}) - \epsilon' > V(s) + \epsilon - \epsilon' = V(s) + \epsilon'$$

We see, that $B((x, s_{-i}), \frac{\epsilon'}{L}) \subset U_{V(s)+\epsilon'}$

This gives:

$$B_i(x, \frac{\epsilon'}{L}) \subset \{x \in S_i : (x, s_{-i}) \in B((x, s_{-i}), \frac{\epsilon'}{L})\}$$

and

$$B_i(x, \frac{\epsilon'}{L}) \subset U_{i,s}$$

Since we have assumed (*) we have lower bound:

$$\mathcal{V}_i(s)(U_{V(s)+\epsilon'}) > \inf\limits_{x \in S_i} \mu_0^i(B_i(x, \frac{\epsilon'}{L})) = \delta_i$$

Thus, we can take $\delta = \min\limits_{i \in P} \alpha_i \delta_i$ and we have required lower bound for f.

Let $\Delta_k = V^{-1}([(k\epsilon', (k+1)\epsilon'))$ for $k \in \mathbb{N}$. Let's further denote:

$$\Delta_{E,k} = \Delta_k \cap \chi_\Gamma^{-1}((-\epsilon, 0]))$$

$$\Delta_{N,k} = \Delta_k \cap \chi_\Gamma^{-1}((-\infty, -\epsilon]))$$

We obviously have following properties:

$$\Delta_k = \Delta_{E,k} \cup \Delta_{N,k}$$

There exists $M \in \mathbb{N}$ s.t.

$$\bigcup_{0 \le k < M} \Delta_k = S$$

and

$$\bigcup_{0 \le k < M} \Delta_{E,k} = \chi_\Gamma^{-1}((-\epsilon, 0]))$$

$$\bigcup_{0 \le k < M} \Delta_{N,k} = \chi_\Gamma^{-1}((-\infty, -\epsilon]))$$

By the definition of the operator A, we have for $j < i$ and $s \in \Delta_i$

$$A(s)(\Delta_j) = 0 \quad (\text{**})$$

For each i we will show:

1. there exists a_k such that $\mu_n(\Delta_k) \xrightarrow[n \to \infty]{} a_k \ge 0$
2. $\mu_n(\Delta_{N,k}) \xrightarrow[n \to \infty]{} 0$

Now we will follow inductive reasoning. First we will prove, that $\mu_n(\Delta_{N,0}) \xrightarrow[n \to \infty]{} 0$ Let's observe, that, because of the property (**):

$$\mu_{n+1}(\Delta_0) = \int_S A(s)(\Delta_0)d\mu_n =$$

$$\int_{\Delta_0} A(s)(\Delta_0)d\mu_n + \int_{S-\Delta_0} A(s)(\Delta_0)d\mu_n = \int_{\Delta_0} A(s)(\Delta_0)d\mu_n \le \mu_n(\Delta_0)$$

So, $\mu_n(\Delta_0)$ is decreasing sequence, thus there is limit $\mu_n(\Delta_0) \xrightarrow[n \to \infty]{} a_0 \ge 0$.

Now, let's assume, that $\mu_n(\Delta_{N,0})$ does not tend to 0 i.e. that there is λ such that for each $M \in \mathbb{N}$ there is $n_M > M$ such that $\mu_{n_M}(\Delta_{N,0}) > \lambda$. Let's take M such that for all $n > M$

$$0 < |\mu_n(\Delta_0) - a_0| = (\mu_n(\Delta_0) - a_0) < \frac{\lambda\delta}{2}$$

Now, let's take $n = n_M > M$ as above. We have:

$$\mu_{n+1}(\Delta_0) = \int_{\Delta_{E,0}} A(s)(\Delta_0)d\mu_n + \int_{\Delta_{N,0}} A(s)(\Delta_0)d\mu_n =$$

$$\int_{\Delta_{E,0}} A(s)(\Delta_0)d\mu_n + \int_{\Delta_{N,0}} A(s)(S - (S - \Delta_0))d\mu_n =$$

$$\int_{\Delta_{E,0}} A(s)(\Delta_0)d\mu_n + \int_{\Delta_{N,0}} A(s)(S)d\mu_n - \int_{\Delta_{N,0}} A(s)(S - \Delta_0)d\mu_n \leq$$

$$\int_{\Delta_{E,0}} A(s)(\Delta_0)d\mu_n + \int_{\Delta_{N,0}} A(s)(S)d\mu_n - \int_{\Delta_{N,0}} A(s)(U_{V(s)+\epsilon'})d\mu_n =$$

$$\int_{\Delta_{E,0}} A(s)(\Delta_0)d\mu_n + \int_{\Delta_{N,0}} A(s)(S)d\mu_n - \int_{\Delta_{N,0}} f d\mu_n <$$

$$\mu_n(\Delta_{E,0}) + \mu_n(\Delta_{N,0}) - \mu_n(\Delta_{N,0})\delta \leq \mu_n(\Delta_0) - \delta\lambda \leq$$

$$a_0 + \frac{\lambda\delta}{2} - \lambda\delta = a_0 - \frac{\lambda\delta}{2}$$

First weak inequality above comes from the fact, that for $s \in \Delta_0$ we have $U_{V(s)+\epsilon'} \subset S - \Delta_0$ This gives contradiction, because

$$\mu_{n+1}(\Delta_0) - a_0 \leq a_0 - \frac{\lambda\delta}{2} - a_0 = -\frac{\lambda\delta}{2}$$

so

$$|\mu_{n+1}(\Delta_0) - a_0| \geq \frac{\lambda\delta}{2}$$

what contradicts our choice of n.

Now, let's assume that assumptions are true for $l \in \{0, \ldots, k-1\}$ and we will show, that they are true also for k. First, let's notice, that as in the case of $k = 0$, because of condition (**) for each n one has:

$$\mu_{n+1}\left(\bigcup_{l\leq k} \Delta_l\right) = \int_S A(s)\left(\bigcup_{l\leq k} \Delta_l\right)d\mu_n =$$

$$\int_{\bigcup_{l\leq k} \Delta_l} A(s)\left(\bigcup_{l\leq k} \Delta_l\right)d\mu_n + \int_{\bigcup_{l>k} \Delta_l} A(s)\left(\bigcup_{l\leq k} \Delta_l\right)d\mu_n =$$

$$\int_{\bigcup_{l\leq k} \Delta_l} A(s)\left(\bigcup_{l\leq k} \Delta_l\right)d\mu_n \leq \mu_n\left(\bigcup_{l\leq k} \Delta_l\right)$$

Thus, $\mu_n(\bigcup_{l\leq k} \Delta_l)$ is decreasing so $\mu_n(\bigcup_{l\leq k} \Delta_l) \xrightarrow[n\to\infty]{} \tilde{a}_0$ for certain $\tilde{a}_k$. Now, since $\mu_n(\bigcup_{l\leq k} \Delta_l) = \sum_{l\leq k} \mu_n(\Delta_l)$ and from induction assumption $\mu_n(\Delta_l)$ converges for $l < k$, so we have that $\mu_n(\Delta_k)$ converges.

Now we can repeat the same reasoning as in the case $k = 0$ to show, that $\mu_n(\Delta_{N,k}) \underset{n\to\infty}{\longrightarrow} 0$.

So, remembering that there is only finite k such that $\Delta_k \neq \emptyset$ we have:

$$1 - \mu_n(NE_{\Gamma,\epsilon}) = \mu_n(S - NE_{\Gamma,\epsilon}) = \mu_n\left(\bigcup_{l\in\mathbb{N}}\Delta_{N,l}\right) = \sum_{l\in\mathbb{N}}\mu_n(\Delta_{N,l}) \underset{n\to\infty}{\longrightarrow} 0$$

so in consequence:

$$\mu_n(NE_{\Gamma,\epsilon}) \underset{n\to\infty}{\longrightarrow} 1$$

We can observe also, that choice of the ϵ above is arbitrary so we can make ϵ arbitrarily small.

5   Comments and Future Work

Theorem 8 makes some relatively strong assumption about metric and measure theoretic properties of the strategy sets S_i and payoff functions u_i. Some of them are essential. E.g. without assumption about lower limit of measures of open balls with constant radius, theorem is not true. It is relatively easy to come up with counterexample taking $S_i = ([0,1][0,1]) \sqcup [0,1]$ being the disjoint union of 2-dimensional cube and interval, where measure as restricted to cube is 2-dimensional Lebesgue (or it's restriction to Borel) and measure of the interval is 0 and preparing V to be small on product of all cubes and high on the product of all intervals. Anyway, in many natural situations assumption is satisfied. E.g. in reasonable case of S_i being the n-dimensional compact convex polytope with n-dimensional Lebesgue measure they are satisfied.

Special choice of the operator A in the Theorem 8 reflects specific random better response dynamics: assumption, that randomly chosen player samples strategy from the set of strategies being better than last strategy responses to strategies played by opponents. Additionally it assumes that the distribution according to which strategy is sampled is "uniform" with respect to measure on the space S_i This assumes the knowledge of the better responses.

More realistic scenario would be following: random player choses strategy sampled from all available to her and verify its outcome. If the outcome is better, she stays with this new strategy. Otherwise she comes back to previously played. In this case we must assume, that no other player will move in the trial play where choice of the strategy is verified, so protocol becomes more complex. Leaving aside this complexity issue, better response dynamics under such trial and error scheme may be still modeled using our formalism. It just requires changing the V_i definitions used to define operator A to following one:

$$\mathcal{V}_i(s)(B) = \begin{cases} \tilde{\mathcal{V}}_i(s)(B \cap U_{i,s}) + \tilde{\mathcal{V}}_i(S_i - U_{i,s}) & s \in B \\ \tilde{\mathcal{V}}_i(s)(B \cap U_{i,s}) & otherwise \end{cases}$$

Theorem 8 for such stochastic better response dynamics should still be valid with essentially the same proof. Also for other distributions on the set of better strategies, as long as conditions similar to (*) are valid for them and operator A remains well defined, should lead to convergence in measure to $NE_{\Gamma,\epsilon}$. However the exact form of such a more general theorem is not yet formulated.

In general, when thinking about stochastic optimization as discussed in the Sect. 2 we treat the cost of every trial in terms of increasing number of steps executed, so a constant. In game-theoretic approach, when searching is done by playing actual game, cost of searching for equilibrium depends on the actual payoff in the game so can be potentially very high and not constant. Analysis of algorithms of reaching equilibrium from this perspective must then consider "cost of searching".

Assumption about lack of collision between players when choosing strategy is essential in the proof of the Theorem 8, since it warranties non-decreasing of the potential V along the trajectories in the space of game profiles. It is an interesting question under what assumption about game and probability of collision one can still assure convergence to equilibrium. In such a case, some special steps have to be built in protocol executed by players to assure, that they are able to distinguish (at least with high probability) between situations when change in the outcome of payoff function is caused by their choice of the strategy and the situation when someone else moved. In such cases communication between players or implementing some "backoff" scheme may play a role.

Comparing to version postulated in [3] the Theorem 8 assumes potential game, not an ordered potential game. This stronger assumption we make seems to be rather technical and it looks like Theorem 8 remains true for ordered potential game, but the proof, even if can follow the lines of the one presented here, would be much more technical.

It is of great interest how one can go beyond assumption that game is potential. It leads to the realm of questions similar to these addressed by Young in [7] under the name of "learning by trial and errors". A part of the planned future studies is a question if function χ_Γ we introduced may be used as a tool in the proofs of theorems and designing protocols of reaching equilibrium in a general game case, reducing it to searching maximum of this function.

A   Appendix

Following is the proof of measurability of the function f defined in the proof of Theorem 8. We will infer it from two lemmas.

Lemma 1. *Let P be the finite set, and for each $i \in P$ S_i be a compact metric space with metric d_i. For each i by σ_i we will denote σ-algebra of Borel sets and μ_i be a regular probabilistic measure on S_i. Let $S = \prod_{i \in P}$ be a metric space with metric d:*

$$d(x,y) = \sum_{i \in P} d_i(x_i, y_i).$$

which gives a product topology in S. We denote by σ the σ-algebra of Borel set in S, and take measure $\mu = \bigotimes_{i \in P} \mu_i$. Let $V : S \to \mathbb{R}$ be continuous, $\alpha \geq 0$ and *function $g_i : S \to \mathbb{R}$ be defined as follows:*

$$g_i(s) = \mu_i(\iota_{i,s}^{-1}(U_{V(s)+\alpha}))$$

where $U_t = V^{-1}((t, \infty))$. Under this assumptions we have: g_i is lower semi-continuous

Proof. We will show, that $g_i^{-1}((\rho, \infty))$ for $\rho \in \mathbb{R}$ is open. Let's assume that $g_i^{-1}((\rho, \infty)) \neq \emptyset$ (if it is empty, it is open trivially) and take $s \in g_i^{-1}((\rho, \infty))$. From definition we have:

$$\mu_i(\iota_{i,s}^{-1}(U_{V(s)+\alpha})) > \rho$$

Since μ_i is regular, there is a compact set K such that $K \subset \iota_{i,s}^{-1}(U_{V(s)+\alpha})$ and $\mu_i > \rho$ Obviously, $\iota_{i,s}(K)$ is compact so V takes it's extrema on K. Thus, there is $\epsilon > 0$ such that, for each $x \in \iota_{i,s}(K)$ we have $V(x) > V(s) + \alpha + \epsilon$ For each $x \in \iota_{i,s}(K)$ then, there is open open neighborhood O_x of x such that: $O_x = \prod_{j \in P} O_x^j$, where O_x^j is neighborhood of x_j in S_j and for each $y \in O_x$ we have $V(y) > V(s) + \alpha + \frac{\epsilon}{2}$ From the compactness, we can over $\iota_{i,s}(K)$ by finite number of these neighborhoods:

$$\iota_{i,s}(K) \subset \bigcup_{j \in \{0, \ldots, n\}} O_{x_j} = \bigcup_{j \in \{0, \ldots, n\}} O_j$$

where last equality is introduced to simplify further referencing these sets. Now let $O_s = V^{-1}((-\infty, V(s) + \frac{\epsilon}{2}))$. It is open neighborhood of s in S. Finally let's construct another open neighborhood of s following way:

$$O = O_s \cap ((\bigcap_{j=0, \ldots, n} O_j^{-i}) \times S_i)$$

where $O_j^{-i} = \prod_{l \in (P - \{i\})} O_j^l$

We will show, that $O \subset g_i^{-1}((\rho, \infty))$.

Let's take $s' \in O$. For each $x \in K$ we have: $\iota_{i,s'}(x) = (x, s'_{-i})$ where $s'_{-i} \in \bigcap_{j=0, \ldots, n} O_j^{-i}$, thus belongs to all O_j^{-i}. Since O_j was cover of $\iota_{i,s}(K)$, there is specific j, such that $(x, s_{-i}) \in O_j$. So, also $(x, s'_{-i}) \in O_j$. Now, we have:

$$V(\iota_{i,s'}(x)) > V(s) + \alpha + \frac{\epsilon}{2} > V(s') + \alpha$$

thus $K \subset \iota_{i,s'}^{-1}(U_{V(s')} + \alpha$ and since $\mu_i(K) > \rho$ we have $O \subset g_i^{-1}((\rho, \infty))$. This completes the proof.

Lemma 2. *Let (X, d) be a metric space, $f : X \to \mathbb{R}$ and $g : X \to \mathbb{R}$ be two lower semi-continuous functions such that, for each $x \in M$ we have $0 \leq g(x) \leq f(x) \leq M$. Let $\phi : X \to \mathbb{R}$ be defined as*

$$\phi(x) = \begin{cases} 1 & g(x) = f(x) \\ \frac{g(x)}{f(x)} & otherwise \end{cases}$$

then ϕ is measurable in the sense of the Borel σ-algebra

Proof. We will show, that ϕ is Baire class 2 function (so measurable with regard to Borel's σ-algebra. Let's take sequence $(\epsilon_n)_{n\in\mathbb{N}}$, such that: $\epsilon_n \xrightarrow[n\to\infty]{} 0$ for each $n \in \mathbb{N}$ we have $\epsilon_n > \epsilon_{n+1}$.

Obviously functions $f_n = f + \epsilon_n$ and $g_n = g + \epsilon_n$ are also lower semi-continuous, and what's more strictly positive ($> \epsilon_n$). From Baire theorem on lower semi-continuous functions [8] there are pointwise increasing sequence of continuous functions $(f_{n,i})_{i\in\mathbb{N}}$ and $(g_{n,i})_{i\in\mathbb{N}}$ such that pointwise $f_{n,i} \xrightarrow[i\to\infty]{} f_n$ and $g_{n,i} \xrightarrow[i\to\infty]{} g_n$. By taking

$$\tilde{f}_{n,i}(x) = max\{f_{n,i}(x), \frac{\epsilon}{2}\}$$

and

$$\tilde{g}_{n,i}(x) = max\{g_{n,i}(x), \frac{\epsilon}{2}\}$$

we can assure, that sequences are bounded below from 0 and taking

$$\tilde{\tilde{f}}_{n,i}(x) = max\{\tilde{f}_{n,i}(x), \tilde{g}_{n,i}(x)\}$$

and

$$\tilde{\tilde{g}}_{n,i}(x) = min\{\tilde{f}_{n,i}(x), \tilde{g}_{n,i}(x)\}$$

we can assure that $\tilde{\tilde{g}}_{n,i} \leq \tilde{\tilde{f}}_{n,i}$. Now, we have $\frac{\tilde{\tilde{g}}_{n,i}}{\tilde{\tilde{f}}_{n,i}} \xrightarrow[i\to\infty]{} \frac{g_n}{f_n}$ so $\frac{g_n}{f_n}$ is of first category as a pointwise limit of continuous functions.

We can also check (by easy calculation), that for each $x \in X$ $\frac{g_n(x)}{f_n(x)} \geq \frac{g_{n+1}(x)}{f_{n+1}(x)}$. Thus $\frac{g_n}{f_n}$ is pointwise convergent equal to 1 whenever $f(x) = g(x)$, so convergent to ϕ. Thus ϕ is of second Baire category.

Let us now recall definition of function f from Theorem 8 (we will use notation and assumptions from Sect. 4 and proof of Theorem 8):

$$f(s) = A(s)(U_{V(s)+\epsilon})$$

We have for each $i \in P$ that functions $f_i(s) = \tilde{\mathcal{V}}_i(s)(U_{i,s}) = \mu_0^i(\iota_{i,s}^{-1}(U_{V(s)}))$ and $g_i(s) = \tilde{\mathcal{V}}_i(s)((U_{V(s)+\epsilon'})) = \mu_0^i(s)(\iota_{i,s}^{-1}(U_{V(s)+\epsilon'}))$ both are lower semi-continuous by Lemma 1, and what's more satisfy conditions of Lemma 2. So, $\mathcal{V}_i(U_{V(s)+\epsilon'})$ is measurable for each i and $f(s) = A(s)(U_{V(s)+\epsilon'}) = \sum_{i\in P} \alpha_i \mathcal{V}_i(U_{V(s)+\epsilon'})$ is measurable as a linear combination of the measurable functions.

References

1. Giry, M.: A categorical approach to probability theory. In: Banaschewski, B. (ed.) Categorical Aspects of Topology and Analysis. LNM, vol. 915, pp. 68–85. Springer, Heidelberg (1982). https://doi.org/10.1007/BFb0092872
2. Gutjahr, W.J.: Stochastic search in metaheuristics. In: Gendreau, M., Potvin, J.Y. (eds.) Handbook of Metaheuristics. ISOR, vol. 146, pp. 573–597. Springer, Boston (2010). https://doi.org/10.1007/978-1-4419-1665-5_19
3. MacKenzie, A.B., DaSilva, L.A.: Game Theory for Wireless Engineers. Synthesis Lectures on Communications. Morgan and Claypool Publishers, San Rafael (2006)
4. Ombach, J., Tarłowski, D.: Nonautonomous stochastic search in global optimization. J. Nonlinear Sci. **22**(2), 169–185 (2012)
5. Solis, F.J., Wets, R.J.B.: Minimization by random search techniques. Math. Oper. Res. **6**(1), 19–30 (1981)
6. Tarłowski, D.: Nonautonomous dynamical systems in stochastic global optimization. Ph.D. thesis, Department of Mathematics, Jagiellonian University (2014)
7. Young, H.P.: Learning by trial and error. Games Econ. Behav. **65**(2), 626–643 (2009)
8. Łojasiewicz, S.: An Introduction to the Theory of Real Functions. Wiley, Hoboken (1988)

Games for Economy and Resource Allocation

Optimal Resource Allocation over Networks via Lottery-Based Mechanisms

Soham Phade$^{(\boxtimes)}$ and Venkat Anantharam

University of California at Berkeley, Berkeley, CA 94720, USA
soham_phade@berkeley.edu, ananth@eecs.berkeley.edu

Abstract. We show that, in a resource allocation problem, the ex ante aggregate utility of players with cumulative-prospect-theoretic preferences can be increased over deterministic allocations by implementing lotteries. We formulate an optimization problem, called the system problem, to find the optimal lottery allocation. The system problem exhibits a two-layer structure comprised of a permutation profile and optimal allocations given the permutation profile. For any fixed permutation profile, we provide a market-based mechanism to find the optimal allocations and prove the existence of equilibrium prices. We show that the system problem has a duality gap, in general, and that the primal problem is NP-hard. We then consider a relaxation of the system problem and derive some qualitative features of the optimal lottery structure.

Keywords: Resource allocation · Networks · Lottery · Cumulative prospect theory

1 Introduction

We consider the problem of congestion management in a network, and resource allocation amongst heterogeneous users, in particular human agents, with varying preferences. This is a well-recognized problem in network economics, with applications to transportation and telecommunication networks, energy smart grids, information and financial networks, labor markets and social networks, to name a few [20]. Market-based solutions have proven to be very useful for this purpose, with varied mechanisms, such as auctions and fixed rate pricing [9]. In this paper, we consider a lottery-based mechanism, as opposed to the deterministic allocations studied in the literature. We mainly ask the following questions: (i) *Do lotteries provide an advantage over deterministic implementations?* (ii) *If yes, then does there exist a market-based mechanism to implement an optimum lottery?*

Research supported by the NSF Science and Technology Center grant CCF- 0939370: "Science of Information", the NSF grants ECCS-1343398, CNS-1527846 and CIF-1618145, and the William and Flora Hewlett Foundation supported Center for Long Term Cybersecurity at Berkeley.

K. Avrachenkov et al. (Eds.): GameNets 2019, LNICST 277, pp. 51–70, 2019.
https://doi.org/10.1007/978-3-030-16989-3_4

In order to answer the first question we need to define our goal in allocating resources. There is an extensive literature on the advantages of lotteries: Eckhoff [7] and Stone [25] hold that lotteries are used because of fairness concerns; Boyce [3] argues that lotteries are effective to reduce rent-seeking from speculators; Morgan [19] shows that lotteries are an effective way of financing public goods through voluntary funds, when the entity raising funds lacks tax power; Hylland and Zeckhauser [10] propose implementing lotteries to elicit honest preferences and allocate jobs efficiently. In all of these works, there is an underlying assumption, which is also one of the key reasons for the use of lotteries, that the goods to be allocated are indivisible.

However, we notice lotteries being implemented even when the goods to be allocated are divisible, for example in lottos and parimutuel betting. In several experiments [22], it has been observed that lottery-based rewards are more appealing than deterministic rewards of the same expected value, and thus provide an advantage in maximizing the desired influence on people's behavior. We also observe several firms presenting lottery-based offers to incentivize customers into buying their products or using their services, and in return to improve their revenues. Thus, although lottery-based mechanisms are being widely implemented, a theoretical understanding for the same seems to be lacking. This is one of the motivations for this paper, which aims to justify the use of lottery-based mechanisms, based on models coming from behavioral economics for how humans evaluate options.

We take a utilitarian approach of maximizing the ex ante aggregate utility or the net happiness of the players. (See [1] and the references therein for the relation with other goals such as maximizing revenue.) We model each player's utility using cumulative prospect theory (CPT), a framework pioneered by Tversky and Kahneman [27], which is believed, based on extensive experimentation with human subjects [4], to form a better theory with which to model human behavior when faced with prospects than is expected utility theory (EUT) [28]. It is important to emphasize that CPT includes EUT as a special case and therefore provides a strict generalization of existing modeling techniques.

CPT posits a *probability weighting function* that, along with the ordering of the allocation outcomes in a lottery, dictates the *probabilistic sensitivity* of a player (details in Sect. 2), a property that plays an important role in lotteries and gambling. As Boyce [3] points out, "it is the lure of getting the good without having to pay for it that gives allocation by lottery its appeal." The probability weighting function typically over-weights small probabilities and under-weights large probabilities, and this captures the "lure" effect. CPT also posits a reference point that divides the prospect outcomes into gains and losses domains in order to model the loss aversion of the players. In order to focus on the effects of probabilistic sensitivity, and to avoid the complications resulting from reference point considerations, we assume that the reference point of all the players is equal to 0, and we consider prospects with only nonnegative outcomes. This is, in fact, identical to the rank dependent utility (RDU) model [23].

We will mainly be concerned with the framework considered in [13], that of throughput control in the internet with elastic traffic. However, this framework is general enough to have applications to network resource allocation problems arising in several other domains. Kelly suggested that the throughput allocation problem be posed as one of achieving maximum aggregate utility for the users. A market is proposed, in which each user submits an amount she is willing to pay per unit time to the network based on tentative rates that she received from the network; the network accepts these submitted amounts and determines the price of each network link. A user is then allocated a throughput in proportion to her submitted amount and inversely proportional to the sum of the prices of the links she wishes to use. Under certain assumptions, Kelly shows that there exist equilibrium prices and throughput allocations, and that these allocations achieve maximum aggregate utility. Thus the overall *system problem* of maximizing aggregate utility is decomposed into a *network problem* and several *user problems*, one for each individual user. Further, in [14], the authors have proposed two classes of algorithms which can be used to implement a relaxation of the above optimization problem.

Instead of allocating a single throughput, we consider allocating a *prospect* of throughputs to each user. Such a prospect consists of a finite set of throughputs and a probability assigned to each of these throughputs, with the interpretation that one of these throughputs would be realized with its corresponding probability (see Sect. 2 for the definition). We then ask the question of finding the optimum allocation profile of prospects, one for each user, comprised of throughputs and associated probabilities for that user, that maximizes the aggregate utility of all the players, and is also *feasible*. An allocation profile of prospects for each user is said to be feasible if it can be implemented, i.e. there exists a probability distribution over feasible throughput allocations whose marginals for each player agree with their allocated prospects.

If all the players have EUT utility with concave utility function, as is typically assumed to model risk-averseness, one can show that there exists a feasible deterministic allocation that achieves the optimum and hence there is no need to consider lotteries. However, if the players' utility is modeled by CPT, then one can improve over the best aggregate utility obtained through deterministic allocations.

For example, Quiggin [24] considers the problem of distributing a fixed amount amongst several homogeneous players with RDU preferences. He concludes that, under certain conditions on players' RDU preferences, the optimum allocation system is a lottery scheme with a few large prizes and a large number of small prizes, and is strictly preferred over distributing the total amount deterministically amongst the players. In Sect. 5, we extend these results to network settings with heterogeneous players.

In Sect. 2, we describe the network model, the lottery structure, and the CPT model of player utility. We formulate an optimization problem, called the system problem, to find the optimum lottery scheme. The solution of such an optimization problem, as explained in Sect. 2, exhibits a layered structure of finding a *permutation profile*, and corresponding feasible throughput allocations. The permutation profile dictates in which order throughput allocations for each player are coupled together for network feasibility purposes. A player's CPT value for her lottery allocation depends only on the order amongst her own throughputs, and not on the coupling with the other players.

Given a permutation profile, the problem of finding optimum feasible throughput allocations is a convex programming problem, which we call the fixed-permutation system problem, and leads to a nice price mechanism. In Sect. 3, we prove the existence of equilibrium prices that decompose the fixed-permutation system problem into a network problem and several user problems, one for each player, as in [13]. The prices can be interpreted as the cost imposed on the players and can be implemented in several forms, such as waiting times in waiting-line auctions or first-come-first-served allocations [2,26], delay or packet loss in the Internet TCP protocol [15,17], efforts or resources invested by players in a contest [6,18], or simply money or reward points.

Finding the optimum permutation profile, on the other hand, is a non-convex problem. In Sect. 4, we study the duality gap in the system problem and consider a relaxation of the system problem by allowing the permutations to be doubly stochastic matrices instead of restricting them to be permutation matrices. We show that strong duality holds in the relaxed system problem and so it has value equal to the dual of the original system problem (Theorem 2). We also consider the problem where link constraints hold in expectation, called the average system problem, and show that strong duality holds in this case and so it has value equal to the relaxed problem. In Sect. 5, we study the average system problem in further detail, and prove a result on the structure of optimal lotteries. Example 2 establishes that the duality gap in the original system problem can be nonzero and Theorem 3 shows that the primal system problem is NP-hard. In Sect. 6, we conclude with some open problems for future research.

2   Model

Consider a network with a set $[m] = \{1, \ldots, m\}$ of *resources* or *links* and a set $[n] = \{1, \ldots, n\}$ of *users* or *players*. Let $c_j > 0$ denote the finite *capacity* of link $j \in [m]$ and let $c := (c_j)_{j \in [m]} \in \mathbb{R}^m$. (All vectors, unless otherwise specified, will be treated as column vectors.) Each user i has a fixed *route* J_i, which is a non-empty subset of $[m]$. Let A be an $n \times m$ matrix, where $A_{ij} = 1$ if link $j \in J_i$, and $A_{ij} = 0$ otherwise. Let $x := (x_i)_{i \in [n]} \in \mathbb{R}_+^n$ denote an *allocation profile* where user i is allocated the throughput $x_i \geq 0$ that flows through the links in the route J_i. We say that an allocation profile x is feasible if it satisfies the capacity constraints of the network, i.e., $A^T x \leq c$, where the inequality is coordinatewise. Let $\mathcal{F}$ denote the set of all feasible allocation profiles. We assume that the network constraints are such that $\mathcal{F}$ is bounded, and hence a polytope.

Instead of allocating a fixed throughput x_i to player $i \in [n]$, we consider allocating her a *lottery* (or a *prospect*)

$$L_i := \{(p_i(1), y_i(1)), \ldots, (p_i(k_i), y_i(k_i))\}, \tag{1}$$

where $y_i(l_i) \geq 0, l_i \in [k_i]$, denotes a throughput and $p_i(l_i), l_i \in [k_i]$, is the probability with which throughput $y_i(l_i)$ is allocated. We assume the lottery to be exhaustive, i.e. $\sum_{l_i \in [k_i]} p_i(l_i) = 1$. (Note that we are allowed to have $p_i(l_i) = 0$ for some values of $l_i \in [k_i]$ and $y_i(l_i^1) = y_i(l_i^2)$ for some $l_i^1, l_i^2 \in [k_i]$.) Let $L = (L_i, i \in [n])$ denote a *lottery profile*, where each player i is allocated lottery L_i.

We now describe the CPT model we use to measure the "utility" or "happiness" derived by each player from her lottery (for more details see [29]). Each player i is associated with a value function $v_i : \mathbb{R}_+ \to \mathbb{R}_+$ that is continuous, differentiable, concave, and strictly increasing, and a probability weighting function $w_i : [0,1] \to [0,1]$ that is continuous, strictly increasing and satisfies $w_i(0) = 0$ and $w_i(1) = 1$.

For the prospect L_i in (1), let $\pi_i : [k_i] \to [k_i]$ be a permutation such that

$$z_i(1) \geq z_i(2) \geq \cdots \geq z_i(k_i), \tag{2}$$

and

$$y_i(l_i) = z_i(\pi_i(l_i)) \text{ for all } l_i \in [k_i]. \tag{3}$$

The prospect L_i can equivalently be written as

$$L_i = \{(\tilde{p}_i(1), z_i(1)); \ldots ; (\tilde{p}_i(k_i), z_i(k_i))\},$$

where $\tilde{p}_i(l_i) := p_i(\pi_i^{-1}(l_i))$ for all $l_i \in [k_i]$. The *CPT value* of prospect L_i for player i is evaluated using the value function $v_i(\cdot)$ and the probability weighting function $w_i(\cdot)$ as follows:

$$V_i(L_i) := \sum_{l_i=1}^{k_i} d_{l_i}(p_i, \pi_i) v_i(z_i(l_i)), \tag{4}$$

where $d_{l_i}(p_i, \pi_i)$ are the *decision weights* given by $d_1(p_i, \pi_i) := w_i(\tilde{p}_i(1))$ and

$$d_{l_i}(p_i, \pi_i) := w_i(\tilde{p}_i(1) + \cdots + \tilde{p}_i(l_i)) - w_i(\tilde{p}_i(1) + \cdots + \tilde{p}_i(l_i - 1)),$$

for $1 < l_i \leq k_i$. Although the expression on the right in Eq. (4) depends on the permutation π_i, one can check that the formula evaluates to the same value $V_i(L_i)$ as long as π_i satisfies (2) and (3). The CPT value of prospect L_i, can equivalently be written as

$$V_i(L_i) = \sum_{l_i=1}^{k_i} w_i \Big(\sum_{s_i=1}^{l_i} \tilde{p}_i(s_i) \Big) [v_i(z_i(l_i)) - v_i(z_i(l_i + 1)))],$$

where $z_i(k_i + 1) := 0$. Thus the lowest allocation $z_i(k_i)$ is weighted by $w_i(1) = 1$, and every increment in the value of the allocations, $v_i(z_i(l_i)) - v_i(z_i(l_i+1)), \forall l_i \in [k_i - 1]$, is weighted by the probability weighting function of the probability of receiving an allocation at least equal to $z_i(l_i)$.

For any finite set S, let $\Delta(S)$ denote the standard simplex of all probability distributions on the set S, i.e.,

$$\Delta(S) := \{(p(s), s \in S) | p(s) \geq 0 \; \forall s \in S, \sum_{s \in S} p(s) = 1\}.$$

Thus $p_i := (p_i(l_i))_{l_i \in [k_i]} \in \Delta([k_i])$. We say that a lottery profile L is *feasible* if there exists a joint distribution $p \in \Delta(\prod_i [k_i])$ such that the following conditions are satisfied:

(i) The marginal distributions agree with L_i for all players i, i.e. $\sum_{l_{-i}} p(l_i, l_{-i}) = p_i(l_i)$ for all $l_i \in [k_i]$, where l_{-i} in the summation ranges over values in $\prod_{i' \neq i} [k_{i'}]$.

(ii) For each $(l_i)_{i \in [n]} \in \prod_i [k_i]$ in the support of the distribution p (i.e. $p((l_i)_{i \in [n]}) > 0$), the allocation profile $(y_i(l_i))_{i \in [n]}$ is feasible.

The distribution p and the throughputs $(y_i(l_i), i \in [n], l_i \in [k_i])$ of a feasible lottery profile together define a *lottery scheme*. In the following, we restrict our attention to specific types of lottery schemes, wherein the network implements with equal probability one of the k allocation profiles $y(l) := (y_i(l))_{i \in [n]} \in \mathbb{R}^n_+$, for $l \in [k]$. Let $[k] = \{1, \ldots, k\}$ denote the set of *outcomes*, where allocation profile $y_i(l)$ is implemented if outcome l occurs. Clearly, such a scheme is feasible iff each of the allocation profiles $y(l), \forall l \in [k]$ belongs to $\mathcal{F}$. Player i thus faces the prospect $L_i = \{(1/k, y_i(l))\}_{l=1}^k$ and such a lottery scheme is completely characterized by the tuple $y := (y_i(l), i \in [n], l \in [k])$. By taking k large enough, any lottery scheme can be approximated by such a scheme.

Let $y_i := (y_i(l))_{l \in [k]} \in \mathbb{R}^k_+$. Let $z_i := (z_i(l))_{l \in [k]} \in \mathbb{R}^k_+$ be a vector and $\pi_i : [k] \to [k]$ be a permutation such that

$$z_i(1) \geq z_i(2) \geq \cdots \geq z_i(k),$$

and

$$y_i(l) = z_i(\pi_i(l)) \text{ for all } l \in [k].$$

Note that y_i is completely characterized by π_i and z_i. Then player i's CPT value will be

$$V_i(L_i) = \sum_{l=1}^k h_i(l) v_i(z_i(l)),$$

where $h_i(l) := w_i(l/k) - w_i((l-1)/k)$ for $l \in [k]$. Let $h_i := (h_i(l))_{l \in [k]} \in \mathbb{R}^k_+$. Note that $h_i(l) > 0$ for all i, l, since the weighting functions are assumed to be strictly increasing.

Looking at the lottery scheme y in terms of individual allocation profiles z_i and permutations π_i for all players $i \in [n]$, allows us to separate those features of y that affect individual preferences and those that pertain to the network implementation. We will later see that the problem of optimizing aggregate utility can be decomposed into two layers: (i) a convex problem that optimizes over resource allocations, and (ii) a non-convex problem that finds the optimal permutation profile.

Let $z := (z_i(l), i \in [n], l \in [k]), \pi := (\pi_i, i \in [n]), h := (h_i(l), i \in [n], l \in [k])$ and $v := (v_i(\cdot), i \in [n])$. Let S_k denote the set of all permutations of $[k]$. The problem of optimizing aggregate utility $\sum_i V_i(L_i)$ subject to the lottery scheme being feasible, can be formulated as follows:

SYS$[z, \pi; h, v, A, c]$

Maximize
$$\sum_{i=1}^{n} \sum_{l=1}^{k} h_i(l) v_i(z_i(l))$$

subject to
$$\sum_{i \in R_j} z_i(\pi_i(l)) \leq c_j, \forall j \in [m], \forall l \in [k],$$

$$z_i(l) \geq z_i(l+1), \forall i \in [n], \forall l \in [k],$$

$$\pi_i \in S_k, \forall i \in [n].$$

Here $R_j := \{i \in [n] | j \in J_i\}$ is the set of all players whose route uses link j. We set $z_i(k+1) = 0$ for all i, and the $z_i(k+1)$ are not treated as variables. This takes care of the condition $z_i(l) \geq 0$ for all $i \in [n], l \in [k]$.

3 Equilibrium

The system problem SYS$[z, \pi; h, v, A, c]$ optimizes over z and π. In this section we fix $\pi_i \in S_k$ for all i and optimize over z. Let us denote this fixed-permutation system problem by SYS_FIX$[z; \pi, h, v, A, c]$.

SYS_FIX$[z; \pi, h, v, A, c]$

Maximize
$$\sum_{i=1}^{n} \sum_{l=1}^{k} h_i(l) v_i(z_i(l))$$

subject to
$$\sum_{i \in R_j} z_i(\pi_i(l)) \leq c_j, \forall j \in [m], \forall l \in [k],$$

$$z_i(l) \geq z_i(l+1), \forall i \in [n], \forall l \in [k].$$

(In contrast with SYS$(z, \pi; \dots)$, in SYS_FIX$(z; \pi, \dots)$, the permutation π is thought of as being fixed.) Since $v_i(\cdot)$ is assumed to be a concave function and $h_i(l) > 0$ for all i, l, this problem has a concave objective function with linear constraints. For all $j \in [m], l \in [k]$, let $\lambda_j(l) \geq 0$ be the dual variables corresponding to the constraints $\sum_{i \in R_j} z_i(\pi_i(l)) \leq c_j$ respectively, and for all $i \in [n], l \in [k]$, let $\alpha_i(l) \geq 0$ be the dual variables corresponding to the constraints $z_i(l) \geq z_i(l+1)$ respectively. Let $\lambda := (\lambda_j(l), j \in [m], l \in [k])$ and $\alpha := (\alpha_i(t), i \in [n], l \in [k])$. Then the Lagrangian for the fixed-permutation system problem SYS_FIX$[z; \pi, h, v, A, c]$ can be written as follows:

$$\mathcal{L}(z;\alpha,\lambda) := \sum_{i=1}^{n}\sum_{l=1}^{k} h_i(l)v_i(z_i(l))$$

$$+ \sum_{i=1}^{n}\sum_{l=1}^{k}\alpha_i(l)[z_i(l) - z_i(l+1)] + \sum_{j=1}^{m}\sum_{l=1}^{k}\lambda_j(l)[c_j - \sum_{i\in R_j} z_i(\pi_i(l))]$$

$$= \sum_{i=1}^{n}\sum_{l=1}^{k}\left[h_i(l)v_i(z_i(l)) + (\alpha_i(l) - \alpha_i(l-1))z_i(l) - \left(\sum_{j\in J_i}\lambda_j(\pi_i^{-1}(l))\right)z_i(l)\right]$$

$$+ \sum_{j=1}^{m}\sum_{l=1}^{k}\lambda_j(l)c_j,$$

where $\alpha_i(0) = 0$ for all $i \in [n]$. Differentiating the Lagrangian with respect to $z_i(l)$ we get,

$$\frac{\partial\mathcal{L}(z;\alpha,\lambda)}{\partial z_i(l)} = h_i(l)v_i'(z_i(l)) + \alpha_i(l) - \alpha_i(l-1) - \left(\sum_{j\in J_i}\lambda_j(\pi_i^{-1}(l))\right).$$

Let

$$\rho_i(l) := \sum_{j\in J_i}\lambda_j(\pi_i^{-1}(l)), \tag{5}$$

for all $i \in [n], l \in [k]$. This can be interpreted as the price per unit throughput for player i for her l-th largest allocation $z_i(l)$. The price of the lottery z_i for player i is given by $\sum_{l=1}^{k}\rho_i(l)z_i(l)$, or equivalently,

$$\sum_{l=1}^{k} r_i(l)\left[z_i(l) - z_i(l+1)\right],$$

where

$$r_i(l) := \sum_{s=1}^{l}\rho_i(s), \text{ for all } l \in [k]. \tag{6}$$

For $l \in [k-1]$, $\alpha_i(l)$ can be interpreted as a transfer of a nonnegative price for player i from her l-th largest allocation to her $(l+1)$-th largest allocation. Since the allocation $z_i(l+1)$ cannot be greater than the allocation $z_i(l)$, there is a subsidy of $\alpha_i(l)$ in the price of $z_i(l)$ and an equal surcharge of $\alpha_i(l)$ in the price of $z_i(l+1)$. This subsidy and surcharge is nonzero (and hence positive) only if the constraint is binding, i.e. $z_i(l) = z_i(l+1)$. On the other hand, $\alpha_i(k)$ is a subsidy in price given to player i for her lowest allocation, since she cannot be charged anything higher than the marginal utility at her zero allocation.

Let $h_i := (h_i(l))_{l \in [k]} \in \mathbb{R}_+^k$. Consider the following user problem for player i:

USER$[m_i; r_i, h_i, v_i]$

Maximize
$$\sum_{l=1}^{k} h_i(l) v_i \left(\sum_{s=l}^{k} \frac{m_i(s)}{r_i(s)} \right) - \sum_{l=1}^{k} m_i(l) \qquad (7)$$

subject to $\quad m_i(l) \geq 0, \forall l \in [k],$

where $r_i := (r_i(l), l \in [k])$ is a vector of rates such that

$$0 < r_i(1) \leq r_i(2) \leq \cdots \leq r_i(k). \qquad (8)$$

We can interpret this as follows: User i is charged rate $r_i(k)$ for her lowest allocation $\delta_i(k) := z_i(k)$. Let $m_i(k)$ denote the budget spent on the lowest allocation and hence $m_i(k) = r_i(k)\delta_i(k)$. For $1 \leq l < k$, she is charged rate $r_i(l)$ for the additional allocation $\delta_i(l) := z_i(l) - z_i(l+1)$, beyond $z_i(l+1)$ up to the next lowest allocation $z_i(l)$. Let $m_i(l)$ denote the budget spent on l-th additional allocation and hence $m_i(l) = r_i(l)\delta_i(l)$.

Let $m := (m_i(l), i \in [n], l \in [k])$ and $\delta := (\delta_i(l), i \in [n], l \in [k])$. Consider the following network problem:

NET$[\delta; m, \pi, A, c]$

Maximize
$$\sum_{i=1}^{n} \sum_{l=1}^{k} m_i(l) \log(\delta_i(l))$$

subject to $\quad \delta_i(l) \geq 0, \forall i, \forall l,$

$$\sum_{i \in R_j} \sum_{s=\pi_i(l)}^{k} \delta_i(s) \leq c_j, \forall j, \forall l.$$

This is the well known Eisenberg-Gale convex program [8] and it can be solved efficiently. Kelly et al. [14] proposed continuous time algorithms for finding equilibrium prices and allocations. For results on polynomial time algorithms for these problems see [5,11]. We have the following decomposition result:

Theorem 1. *For any fixed π, there exist equilibrium parameters r^*, m^*, δ^* and z^* such that*

(i) for each player i, m_i^* solves the user problem USER$[m_i; r_i^*, h_i, v_i]$,
(ii) δ^* solves the network problem NET$[\delta; m^*, \pi, A, c]$,
(iii) $m_i^*(l) = \delta_i^*(l) r_i^*(l)$ for all i, l,
(iv) $\delta_i^*(l) = z_i^*(l) - z_i^*(l+1)$ for all i, l, and
(v) z^* solves the fixed-permutation system problem SYS_FIX$[z; \pi, h, v, A, c]$.

For a proof of this theorem, refer to the ArXiv document [21]. Thus the fixed-permutation system problem can be decomposed into user problems – one for each player – and a network problem, for any fixed permutation profile π. Similar to the framework in [14], we have an iterative process as follows:

The network presents each user i with a rate vector r_i. Each user solves the user problem USER$[m_i; r_i, h_i, v_i]$, and submits their budget vector m_i, The network collects these budget vectors $(m_i)_{i \in [n]}$ and solves the network problem NET$[\delta; m^*, \pi, A, c]$ to get the corresponding allocation z (which can be computed from the incremental allocations δ) and the dual variables λ. The network then computes the rate vectors corresponding to each user from these dual variables as given by (5) and (6) and presents it to the users as updated rates. Theorem 1 shows that the fixed-permutation system problem of maximizing the aggregate utility is solved at the equilibrium of the above iterative process. If the value functions $v_i(\cdot)$ are strictly concave, then one can show that the optimal lottery allocation z^* for the fixed-permutation system problem is unique. However, the dual variables λ, and hence the rates $r_i, \forall i$, need not be unique. Nonetheless, if one uses the continuous-time algorithm proposed in [14] to solve the network problem, then a similar analysis as in [14], based on Lyapunov stability, shows that the above iterative process converges to the equilibrium lottery allocation z^*.

One of the permutation profiles, say π^*, solves the system problem. In the next section, we explore this in more detail. However, it is interesting to note that, for any fixed permutation profile π, any deterministic solution is a special case of the lottery scheme y with permutation profile π. Thus, it is guaranteed that the solution of the fixed-permutation system problem for any permutation profile π is at least as good as any deterministic allocation. Here is a simple example, where a lottery-based allocation leads to strict improvement over deterministic allocations.

Example 1. Consider a network with n players and a single link with capacity c. Let $n = 10$ and $c = 10$. For all players i, we employ the value functions and weighting functions suggested by Kahneman and Tversky [27], given by

$$v_i(x_i) = x_i^{\beta_i}, \beta_i \in [0, 1],$$

and

$$w_i(p_i) = \frac{p_i^{\gamma_i}}{(p^{\gamma_i} + (1-p)^{\gamma_i})^{1/\gamma_i}}, \gamma_i \in (0, 1],$$

respectively. We take $\beta_i = 0.88$ and $\gamma_i = 0.61$ for all $i \in [n]$. These parameters were reported as the best fits to the empirical data in [27]. The probability weighting function is displayed in Fig. 1.

By symmetry and concavity of the value function $v_i(\cdot)$, the optimal deterministic allocation is given by allocating c/n to each player i. The aggregate utility for this allocation is $n * v_1(c/n) = 10$.

Now consider the following lottery allocation: Let $k = n = 10$. Let $\pi_i(l) - 1 = l + i \pmod{k}$ for all $i \in [n]$ and $l \in [k]$. Let $x \in [c/n, c]$ and $z_i(1) = x$ for all $i \in [n]$ and $z_i(l) = (c - x)/(n - 1)$ for all $i \in [n]$ and $l = 2, \ldots, k$. Note that this is a feasible lottery allocation. Such a lottery scheme can be interpreted as follows: Select a "winning" player uniformly at random from all the players. Allocate her

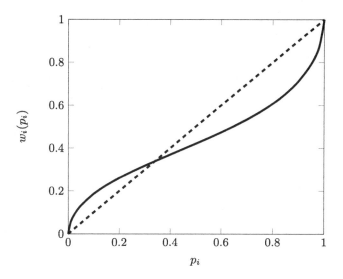

Fig. 1. Probability weighting function

a reward x and equally distribute the remaining reward $c - x$ amongst the rest of the players. The ex ante aggregate utility is given by

$$n * [w_1(1/n)v_1(x) + (1 - w_1(1/n))v_1((c - x)/(n - 1))].$$

This function achieves its maximum equal to 14.1690 at $x = 9.7871$. Thus, the above proposed lottery improves the aggregate utility over any deterministic allocation. The optimum lottery allocation is at least as good as 14.1690.

4   Optimum Permutation Profile and Duality Gap

The system problem $\mathrm{SYS}[z, \pi; h, v, A, c]$ can equivalently be formulated as

$$\max_{\substack{\pi_i \in S_k \forall i, \\ z: z_i(l) \geq z_i(l+1) \forall i, l}} \min_{\lambda \geq 0} \sum_{i=1}^{n} \sum_{l=1}^{k} h_i(l)v_i(z_i(l))$$

$$+ \sum_{j=1}^{m} \sum_{l=1}^{k} \lambda_j(l) \left[c_j - \sum_{i \in R_j} z_i(\pi_i(l)) \right]. \tag{I}$$

Let W_{ps} denote the value of this problem. It is equal to the optimum value of the system problem $\mathrm{SYS}[z, \pi; h, v, A, c]$. By interchanging the max and min, we obtain the following dual problem:

$$\min_{\lambda \geq 0} \quad \max_{\substack{\pi_i \in S_k \forall i, \\ z: z_i(l) \geq z_i(l+1) \forall i, l}} \quad \sum_{i=1}^{n} \sum_{l=1}^{k} h_i(l) v_i(z_i(l))$$

$$+ \sum_{j=1}^{m} \sum_{l=1}^{k} \lambda_j(l) \left[c_j - \sum_{i \in R_j} z_i(\pi_i(l)) \right]. \tag{II}$$

Let W_{ds} denote the value of this dual problem. By weak duality, we know that $W_{ps} \leq W_{ds}$. For a fixed $\lambda \geq 0$ and a fixed z that satisfies $z_i(l) \geq z_i(l+1), \forall i, l$, the optimum permutation profile π in the dual problem (II) should minimize

$$\sum_{j=1}^{m} \sum_{l=1}^{k} \lambda_j(l) \sum_{i \in R_j} z_i(\pi_i(l)),$$

which equals

$$\sum_{i} \sum_{l} \hat{\rho}_i(l) z_i(\pi_i(l)),$$

Here $\hat{\rho}_i(l) := \sum_{j \in J_i} \lambda_j(l)$, is the price per unit allocation for player i under outcome l. Since the numbers $z_i(l)$ are ordered in descending order, any optimal permutation π_i must satisfy

$$\hat{\rho}_i(\pi_i^{-1}(1)) \leq \hat{\rho}_i(\pi_i^{-1}(2)) \leq \cdots \leq \hat{\rho}_i(\pi_i^{-1}(k)). \tag{9}$$

In other words, any optimal permutation profile π of the dual problem (II) must allocate throughputs in the order opposite to that of the prices $\hat{\rho}_i(l)$. We now state a lemma whose proof can be found in the ArXiv document [21].

Lemma 1. *If strong duality holds between the problems* (I) *and* (II), *then any optimum permutation profile* π^* *satisfies* (9) *for all* i.

In general, there is a non-zero duality gap between the problems (I) and (II) (see Example 2 for such an example where the optimum permutation profile π^* does not satisfy (9)).

The permutation π_i can be represented by a $k \times k$ permutation matrix M_i, where $M_i(s,t) = 1$ if $\pi_i(s) = t$ and $M_i(s,t) = 0$ otherwise, for $s,t \in [k]$. The network constraints $\sum_{i \in R_j} z_i(\pi_i(l)) \leq c_j, \forall l \in [k]$, can equivalently be written as $\sum_{i \in R_j} M_i z_i \leq c_j \mathbf{1}$, where $\mathbf{1}$ denotes a vector of appropriate size with all its elements equal to 1, and the inequality is coordinatewise. A possible relaxation of the system problem is to consider doubly stochastic matrices M_i instead of restricting them to be permutation matrices. A matrix is said to be doubly stochastic if all its entries are nonnegative and each row and column sums up to 1. A permutation matrix is hence a doubly stochastic matrix. Let Ω_k denote the set of all doubly stochastic $k \times k$ matrices and let Ω_k^* denote the set of all $k \times k$ permutation matrices.

Let $M = (M_i, i \in [n])$ denote a profile of doubly stochastic matrices. The relaxed system problem can then be written as follows:

SYS_REL$[z, M; h, v, A, c]$

Maximize $$\sum_{i=1}^{n} \sum_{l=1}^{k} h_i(l) v_i(z_i(l))$$

subject to $$\sum_{i \in R_j} M_i z_i \leq c_j \mathbf{1}, \forall j,$$

$$z_i(l) \geq z_i(l+1), \forall i, \forall l,$$

$$M_i \in \Omega_k, \forall i.$$

Then the corresponding primal problem can be written as follows:

$$\max_{\substack{M_i \in \Omega_k \forall i, \\ z: z_i(l) \geq z_i(l+1) \forall l, \forall i}} \min_{\lambda_j \geq 0, \forall j} \sum_i \sum_l h_i(l) v_i(z_i(l))$$

$$+ \sum_j \lambda_j^T \left[c_j \mathbf{1} - \sum_{i \in R_j} M_i z_i \right]. \tag{III}$$

where $\lambda_j = (\lambda_j(l))_{l \in [k]} \in \mathbb{R}_+^k$. Let W_{pr} denote the value of this problem. Interchanging min and max we get the corresponding dual:

$$\min_{\lambda_j \geq 0, \forall j} \max_{\substack{M_i \in \Omega_k \forall i, \\ z: z_i(l) \geq z_i(l+1) \forall l, \forall i}} \sum_i \sum_l h_i(l) v_i(z_i(l))$$

$$+ \sum_j \lambda_j^T \left[c_j \mathbf{1} - \sum_{i \in R_j} M_i z_i \right]. \tag{IV}$$

Let W_{dr} denote the value of this problem. If the link constraints in the relaxed system problem hold then

$$\frac{1}{k} \sum_{i \in R_j} \sum_{l=1}^{k} z_i(l) = \frac{1}{k} \sum_{i \in R_j} \mathbf{1}^T M_i z_i \leq \frac{1}{k} \mathbf{1}^T c_j \mathbf{1} = c_j. \tag{10}$$

This inequality essentially says that the link constraints should hold in expectation. Thus we have the following average system problem:

SYS_AVG$[z; h, v, A, c]$

Maximize $$\sum_{i=1}^{n} \sum_{l=1}^{k} h_i(l) v_i(z_i(l))$$

subject to $$\sum_{i \in R_j} \frac{1}{k} \sum_{l=1}^{k} z_i(l) \leq c_j, \forall j,$$

$$z_i(l) \geq z_i(l+1), \forall i, \forall l,$$

with its corresponding primal problem:

$$\max_{z:z_i(l)\geq z_i(l+1)\forall l,\forall i} \quad \min_{\bar{\lambda}_j\geq 0,\forall j} \quad \sum_i \sum_l h_i(l)v_i(z_i(l))$$
$$+ \sum_j \bar{\lambda}_j \left[c_j - \sum_{i\in R_j} \frac{1}{k} \sum_{l=1}^{k} z_i(l) \right], \tag{V}$$

and the dual problem:

$$\min_{\bar{\lambda}_j\geq 0,\forall j} \quad \max_{z:z_i(l)\geq z_i(l+1)\forall l,\forall i} \quad \sum_i \sum_l h_i(l)v_i(z_i(l))$$
$$+ \sum_j \bar{\lambda}_j \left[c_j - \sum_{i\in R_j} \frac{1}{k} \sum_{l=1}^{k} z_i(l) \right], \tag{VI}$$

where $\bar{\lambda}_j \in \mathbb{R}$ are the dual variables corresponding to the link constraints. Let W_{pa} and W_{da} denote the values of these primal and dual problems respectively.
Then we have the following relation:

Theorem 2. *For any system problem defined by h, v, A and c, we have*

$$W_{ps} \leq W_{pr} = W_{pa} = W_{da} = W_{dr} = W_{ds}.$$

See the ArXiv document [21] for a proof of this theorem. The duality gap is a manifestation of the "hard" link constraints. The relaxed problem is "equivalent" to the average problem and strong duality holds for this relaxation. We will later study the average problem in further detail (Sect. 5).

We observed earlier in Lemma 1 that if strong duality holds in the system problem, then the optimum permutation profile π^* satisfies (9). Consider a simple example of two players sharing a single link. Suppose that, at the optimum, $\lambda(l)$ are the prices for $l \in [k]$ corresponding to this link under the different outcomes, and suppose not all of these are equal. Then the optimum permutation profile of the dual problem will align both players' allocations in the same order, i.e. the high allocations of player 1 will be aligned with the high allocations of player 2. However, we can directly see from the system problem that an optimum π^* should align the two players' allocations in opposite order. The following example builds on this observation and shows that strong duality need not hold for the system problem.

Example 2. Consider the following example with two players $\{1, 2\}$ and a single link with capacity 2.9. Let $k = 2$. Let the corresponding CPT characteristics of the two players be as follows:

$$h_1(1) = \frac{1}{3}, \qquad\qquad h_1(2) = \frac{2}{3},$$
$$h_2(1) = \frac{5}{6}, \qquad\qquad h_2(2) = \frac{1}{6},$$
$$v_1(x) = \log(x + 0.05) + 3, \qquad v_2(x) = \frac{2\log(x + 0.05) + 3(x + 0.05)}{5} + 3.$$

For this problem, it is easy to see that $\pi_1 = (1,2)$ and $\pi_2 = (2,1)$ is an optimal permutation. Solving the fixed-permutation system problem with respect to this permutation we get optimal value equal to 7.5621. The corresponding variable values are

$$z_1(1) = y_1(1) = 1.95, \qquad\qquad z_1(2) = y_1(2) = 0.95,$$
$$z_2(1) = y_2(2) = 1.95, \qquad\qquad z_2(2) = y_2(1) = 0.95,$$

and the dual variable values are

$$\lambda_1(1) = \frac{1}{6}, \qquad\qquad\qquad \lambda_1(2) = \frac{2}{3},$$

and $\alpha_i(l) = 0$, for $i = 1, 2, l = 1, 2$. One can check that these satisfy the KKT conditions.

Let us now evaluate the value of the dual problem (II). By symmetry, we can assume without loss of generality that $\lambda_1(1) \leq \lambda_1(2)$. As a result, optimal permutations for the dual problem are given by $\pi_1 = \pi_2 = (1,2)$. For fixed $\lambda_1(1)$ and $\lambda_1(2)$, we solve the following optimization problem:

$$\max_{\substack{z_1(1) \geq z_1(2) \geq 0 \\ z_2(1) \geq z_2(2) \geq 0}} \quad \frac{1}{3}\log(z_1(1) + 0.05) + \frac{2}{3}\log(z_1(2) + 0.05)$$

$$+ \frac{5}{6}\left[\frac{2\log(z_2(1) + 0.05) + 3(z_2(1) + 0.05)}{5}\right] \qquad\qquad \text{(VII)}$$

$$+ \frac{1}{6}\left[\frac{2\log(z_2(2) + 0.05) + 3(z_2(2) + 0.05)}{5}\right]$$

$$- \lambda_1(1)[z_1(1) + z_2(1)] - \lambda_1(2)[z_1(2) + z_2(2)]$$

$$+ 2.9[\lambda_1(1) + \lambda_1(2)] + 6.$$

If $\lambda_1(1) \leq 0.5$, then the value of the problem (VII) is equal to ∞ (let $z_2(1) \to \infty$). If $\lambda_1(1) > 0.5$ (and hence $\lambda_1(2) > 0.5$ because $\lambda_1(2) \geq \lambda_1(2)$), then we observe that the effective domain of maximization in the problem (VII) is compact and problem (VII) has a finite value. Hence it is enough to consider $\lambda_1(1) > 0.5$. At the optimum there exist $\alpha_1(1), \alpha_1(2), \alpha_2(1), \alpha_2(2) \geq 0$ such that

$$\lambda_1(1) = \frac{1}{3}\frac{1}{z_1(1) + 0.05} + \alpha_1(1),$$

$$\lambda_1(2) = \frac{2}{3}\frac{1}{z_1(2) + 0.05} - \alpha_1(1) + \alpha_1(2),$$

$$\lambda_1(1) = \frac{1}{3}\frac{1}{z_2(1) + 0.05} + \frac{1}{2} + \alpha_2(1),$$

$$\lambda_1(2) = \frac{1}{15}\frac{1}{z_2(2) + 0.05} + \frac{1}{10} - \alpha_2(1) + \alpha_2(2),$$

and

$$\alpha_1(1)[z_1(1) - z_1(2)] = 0, \qquad\qquad \alpha_1(2)z_1(2) = 0,$$
$$\alpha_2(1)[z_2(1) - z_2(2)] = 0, \qquad\qquad \alpha_2(2)z_2(2) = 0.$$

We now consider each of the sixteen (4×4) cases based on whether the inequalities $z_i(l) \geq z_i(l+1)$ for $i = 1, 2$ and $l = 1, 2$, hold strictly or not. Minimizing over feasible pairs $(\lambda_1(1), \lambda_1(2))$ for each of these cases, we get that the minimum value of the objective function is 8.2757 over these cases. (See the ArXiv document [21] for full details.) Thus the optimal dual value is 8.2757 and this is strictly greater than the primal value.

We have the following theorem whose proof can be found in the ArXiv document [21].

Theorem 3. *The primal problem* (I) *is NP-hard.*

5   Average System Problem and Optimal Lottery Structure

Suppose it is enough to ensure that the link constraints are satisfied in expectation, as in the average system problem. Consider the function $V_i^{\text{avg}}(\bar{z}_i)$ on $\mathbb{R}_+$ given by the value of the following optimization problem:

$$\text{Maximize} \quad \sum_{l=1}^{k} h_i(l) v_i(z_i(l))$$

$$\text{subject to} \quad \frac{1}{k} \sum_{l=1}^{k} z_i(l) = \bar{z}_i, \tag{VIII}$$

$$z_i(l) \geq z_i(l+1), \forall l \in [k].$$

Let $Z_i(\bar{z}_i)$ denote the set of feasible $(z_i(l))_{l \in [k]}$ in the above problem for any fixed $\bar{z}_i \geq 0$. We observe that $Z_i(\bar{z}_i)$ is a polytope, and hence $V_i^{\text{avg}}(\bar{z}_i)$ is well defined.

Lemma 2. *For any continuous, differentiable, concave and strictly increasing value function $v_i(\cdot)$, the function $V_i^{avg}(\cdot)$ is continuous, differentiable, concave and strictly increasing in $\bar{z}_i$.*

For a proof of this lemma, refer to the ArXiv document [21]. The average system problem SYS_AVG$[z; h, v, A, c]$ can thus be written as

$$\text{Maximize} \quad \sum_{i=1}^{n} V_i^{\text{avg}}(\bar{z}_i)$$

$$\text{subject to} \quad \sum_{i \in R_j} \bar{z}_i \leq c_j, \forall j,$$

$$\bar{z}_i \geq 0, \forall i.$$

Kelly [13] showed that this problem can be decomposed into user problems, one for each user i,

$$\text{Maximize} \quad V_i^{\text{avg}}(\bar{z}_i) - \bar{\rho}_i \bar{z}_i$$

$$\text{subject to} \quad \bar{z}_i \geq 0,$$

and a network problem

$$\text{Maximize} \quad \sum_{i=1}^{n} \bar{\rho}_i \bar{z}_i$$

$$\text{subject to} \quad \sum_{i \in R_j} \bar{z}_i \leq c_j, \forall j,$$

$$\bar{z}_i \geq 0, \forall i,$$

in the sense that there exist $\bar{\rho}_i \geq 0, \forall i \in [n]$, such that the optimum solutions $\bar{z}_i$ of the user problems, for each i, solve the network problem and the average system problem. Note that this decomposition is different from the one presented in Sect. 3. Here the network problem aims at maximizing its total revenue $\sum_{i=1}^{n} \bar{\rho}_i \bar{z}_i$, instead of maximizing a weighted aggregate utility where the utility is replaced with a proxy logarithmic function. The above decomposition is not as useful as the decomposition in Sect. 3 in order to develop iterative schemes that converge to equilibrium. However, the above decomposition motivates the following user problem:

USER_AVG$[z_i; \bar{\rho}_i, h_i, v_i]$

$$\text{Maximize} \quad \sum_{l=1}^{k} h_i(l) v_i(z_i(l)) - \frac{\bar{\rho}_i}{k} \sum_{l=1}^{k} z_i(l)$$

$$\text{subject to} \quad z_i(l) \geq z_i(l+1), \forall l \in [k],$$

where, as before, $z_i(k+1) = 0$.

We observed in Theorem 2 that strong duality holds in the average system problem. Let z^* be the optimum lottery scheme that solves this problem. Then, first of all, z^* satisfies $z_i^*(l) \geq z_i^*(l+1)$ $\forall i, l$ and is feasible in expectation, i.e., $\bar{z}^* := (\bar{z}_i^*)_{i \in [n]} \in \mathcal{F}$, where $\bar{z}_i^* := (1/k) \sum_l z_i^*(l)$. Further, z^* optimizes the objective function of the average system problem. Besides, there exist $\bar{\lambda}_j^* \geq 0$ for all j such that the primal average problem (V) and the dual average problem (VI) each attain their optimum at $\{z^*, (\bar{\lambda}_j^*, j \in [m])\}$.

For player i, consider the price $\bar{\rho}_i^* := \sum_{j \in J_i} \bar{\lambda}_j^*$, which is obtained by summing the prices $\bar{\lambda}_j^*$ corresponding to the links on player i's route. From the dual average problem (VI), fixing $\bar{\lambda}_j = \bar{\lambda}_j^*$ $\forall j$, we get that the optimum lottery allocation z_i^* for player i should optimize the problem USER_AVG$[z_i; \bar{\rho}_i^*, h_i, v_i]$.

We now impose some additional conditions on the probability weighting function that are typically assumed based on empirical evidence and certain psychological arguments [12]. We assume that the probability weighting function $w_i(p_i)$ is concave for small values of the probability p_i and convex for the rest. Formally, there exists a probability $\tilde{p}_i \in [0,1]$ such that $w_i(p_i)$ is concave over the interval $p_i \in [0, \tilde{p}_i]$ and convex over the interval $[\tilde{p}_i, 1]$. Typically the point of inflection, $\tilde{p}_i$, is around $1/3$.

Let $w_i^* : [0,1] \to [0,1]$ be the minimum concave function that dominates $w_i(\cdot)$, i.e., $w_i^*(p_i) \geq w_i(p_i)$ for all $p_i \in [0,1]$. Let $p^* \in [0,1]$ be the smallest probability such that $w_i^*(p_i)$ is linear over the interval $[p_i^*, 1]$. We now state a lemma whose proof is included in the ArXiv document [21].

Lemma 3. *Given the assumptions on $w_i(\cdot)$, we have $p_i^* \leq \tilde{p}_i$ and $w_i^*(p_i) = w_i(p_i)$ for $p_i \in [0, p_i^*]$. If $p_i^* < 1$, then for any $p_i^1 \in [p_i^*, 1)$, we have*

$$w_i(p_i) \leq w_i(p_i^1) + (p_i - p_i^1)\frac{1 - w_i(p_i^1)}{1 - p_i^1}. \tag{11}$$

for all $p_i \in [p_i^1, 1]$.

We now show that, under certain conditions, the optimal lottery allocation z_i^* satisfies

$$z_i^*(l^*) = z_i^*(l^* + 1) = \cdots = z_i^*(k), \tag{12}$$

where $l^* := \min\{l \in [k] : (l-1)/k \geq p_i^*\}$, provided $p_i^* \leq (k-1)/k$. As a result, for a typical optimum lottery allocation, the lowest allocation occurs with a large probability approximately equal to $1 - p_i^*$, and with a few higher allocations that we recognize as bonuses. We have the following proposition whose proof can be found in the ArXiv document [21].

Proposition 1. *For any average user problem $USER\_AVG[z_i; \bar{\rho}_i^*, h_i, v_i]$ with a strictly increasing, continuous, differentiable and strictly concave value function $v_i(\cdot)$, and a strictly increasing continuous probability weighting function $w_i(\cdot)$ (satisfying $w_i(0) = 0$ and $w_i(1) = 1$) such that $p_i^* \leq (k-1)/k$, the optimum lottery allocation z_i^* satisfies Eq. (12).*

6   Conclusions and Future Work

We saw that if we take the probabilistic sensitivity of players into account, then lottery allocation improves the ex ante aggregate utility of the players. We considered the RDU model, a special case of CPT utility, to model probabilistic sensitivity. This model, however, is restricted to reward allocations, and it would be interesting to extend it to a general CPT model with reference point and loss aversion. This will allow us to study loss allocations as in punishment or burden allocations, for example criminal justice, military drafting, etc.

For any fixed permutation profile, we showed the existence of equilibrium prices in a market-based mechanism to implement an optimal lottery. We also saw that finding the optimal permutation profile is an NP-hard problem. We note that the system problem has parallels in cross-layer optimization in wireless [16] and multi-route networks [30]. Several heuristic methods have helped achieve approximately optimal solutions in cross-layer optimization. Similar methods need to be developed for our system problem. We leave this for future work.

The hardness in the system problem comes from hard link constraints. Hence, by relaxing these conditions to hold only in expectation, we derived some qualitative features of the optimal lottery structure under the typical assumptions on the probability weighting function of each agent in the RDU model. As observed, the players typically ensure their minimum allocation with high probability, and gamble for higher rewards with low probability.

References

1. Altman, E., Wynter, L.: Equilibrium, games, and pricing in transportation and telecommunication networks. Netw. Spatial Econ. **4**(1), 7–21 (2004)
2. Barzel, Y.: A theory of rationing by waiting. J. Law Econ. **17**(1), 73–95 (1974)
3. Boyce, J.R.: Allocation of goods by lottery. Econ. Inquiry **32**(3), 457–476 (1994)
4. Camerer, C.F.: Prospect theory in the wild: evidence from the field. In: Choices, Values, and Frames. pp. 288–300. Contemporary Psychology. No. 47. American Psychology Association, Washington, DC (2001)
5. Chakrabarty, D., Devanur, N., Vazirani, V.V.: New results on rationality and strongly polynomial time solvability in Eisenberg-Gale markets. In: Spirakis, P., Mavronicolas, M., Kontogiannis, S. (eds.) WINE 2006. LNCS, vol. 4286, pp. 239–250. Springer, Heidelberg (2006). https://doi.org/10.1007/11944874_22
6. Che, Y.K., Gale, I.: Optimal design of research contests. Am. Econ. Rev. **93**(3), 646–671 (2003)
7. Eckhoff, T.: Lotteries in allocative situations. Inf. (Int. Soc. Sci. Council) **28**(1), 5–22 (1989)
8. Eisenberg, E., Gale, D.: Consensus of subjective probabilities: the pari-mutuel method. Ann. Math. Stat. **30**(1), 165–168 (1959)
9. Falkner, M., Devetsikiotis, M., Lambadaris, I.: An overview of pricing concepts for broadband IP networks. IEEE Commun. Surv. Tutorials **3**(2), 2–13 (2000)
10. Hylland, A., Zeckhauser, R.: The efficient allocation of individuals to positions. J. Polit. Econ. **87**(2), 293–314 (1979)
11. Jain, K., Vazirani, V.V.: Eisenberg-Gale markets: algorithms and game-theoretic properties. Games Econ. Behav. **70**(1), 84–106 (2010)
12. Kahneman, D., Tversky, A.: Prospect theory: an analysis of decision under risk. Econometrica **47**(2), 263–292 (1979)
13. Kelly, F.: Charging and rate control for elastic traffic. Eur. Trans. Telecommun. **8**(1), 33–37 (1997)
14. Kelly, F.P., Maulloo, A.K., Tan, D.K.: Rate control for communication networks: shadow prices, proportional fairness and stability. J. Oper. Res. Soc. **49**(3), 237–252 (1998)
15. La, R.J., Anantharam, V.: Utility-based rate control in the internet for elastic traffic. IEEE/ACM Trans. Netw. (TON) **10**(2), 272–286 (2002)
16. Lin, X., Shroff, N.B., Srikant, R.: A tutorial on cross-layer optimization in wireless networks. IEEE J. Sel. Areas Commun. **24**(8), 1452–1463 (2006)
17. Mo, J., Walrand, J.: Fair end-to-end window-based congestion control. IEEE/ACM Trans. Netw. **8**(5), 556–567 (2000)
18. Moldovanu, B., Sela, A.: The optimal allocation of prizes in contests. Am. Econ. Rev. **91**(3), 542–558 (2001)
19. Morgan, J.: Financing public goods by means of lotteries. Rev. Econ. Stud. **67**(4), 761–784 (2000)
20. Nagurney, A.: Network Economics: A Variational Inequality Approach, vol. 10. Springer, Heidelberg (2013)
21. Phade, S.R., Anantharam, V.: Optimal resource allocation over networks via lottery-based mechanisms. arXiv preprint (2018)
22. Prabhakar, B.: Designing large-scale nudge engines. In: ACM SIGMETRICS Performance Evaluation Review, vol. 41, pp. 1–2. ACM (2013)
23. Quiggin, J.: A theory of anticipated utility. J. Econ. Behav. Organ. **3**(4), 323–343 (1982)

24. Quiggin, J.: On the optimal design of lotteries. Economica **58**(229), 1–16 (1991)
25. Stone, P., Political: Why lotteries are just. J. Polit. Philos. **15**(3), 276–295 (2007)
26. Taylor, G.A., Tsui, K.K., Zhu, L.: Lottery or waiting-line auction? J. Public Econ. **87**(5–6), 1313–1334 (2003)
27. Tversky, A., Kahneman, D.: Advances in prospect theory: cumulative representation of uncertainty. J. Risk Uncertainty **5**(4), 297–323 (1992)
28. Von Neumann, J., Morgenstern, O.: Theory of games and economic behavior. Bull. Am. Math. Soc. **51**(7), 498–504 (1945)
29. Wakker, P.P.: Prospect Theory: For Risk and Ambiguity. Cambridge University Press, Cambridge (2010)
30. Wang, J., Li, L., Low, S.H., Doyle, J.C.: Cross-layer optimization in TCP/IP networks. IEEE/ACM Trans. Netw. (TON) **13**(3), 582–595 (2005)

Two-Level Cooperation in Network Games

Leon Petrosyan⬤ and Artem Sedakov$^{(\boxtimes)}$⬤

Saint Petersburg State University, 7/9 Universitetskaya nab.,
Saint Petersburg 199034, Russia
{l.petrosyan,a.sedakov}@spbu.ru

Abstract. The problem of allocating a value in hierarchical cooperative structures is important in the game theoretic literature, and it often arises in practice. In this paper, we consider a two-level structure of players communication and propose a procedure allocating the value in two steps: first the value is allocated at the upper level among groups of players, and then each group allocates the designated value among its members. We demonstrate how to allocate the value in two steps using the Shapley value and show the difference with the classical one-step allocation procedure. We then adopt this approach for games with pairwise interactions and provide relations between several definitions of the characteristic function and the corresponding Shapley values.

Keywords: Network · Hierarchy · Cooperation ·
Two-level allocation · Shapley value

1 Introduction

Networks are very natural and convenient in describing and representing hierarchical communication structures. Such hierarchies model situations in which players (agents) are at different level of subordination. The simplest examples of the usage of hierarchies include leader-follower models, organization structures indicating communication and subordination between departments. When the hierarchical communication structure is specified, one can study the problem of optimal behavior of players. The game theoretic literature covers two types of behavior: equilibrium and cooperative. In this paper, we focus on a cooperative case as it can provide players with a better outcome in the game rather than an equilibrium outcome; however, the cooperative model will be built by using an initially given non-cooperative game. As it is common in the cooperative game theory, players jointly choose their actions to achieve the largest total payoff followed by its allocation among them. The hierarchical communication structure naturally dictates that this value has to be allocated step-by-step from

This research was supported by the Russian Science Foundation (grant No. 17-11-01079).

K. Avrachenkov et al. (Eds.): GameNets 2019, LNICST 277, pp. 71–81, 2019.
https://doi.org/10.1007/978-3-030-16989-3_5

the topmost level of hierarchy to its lowermost level. Specifically, in this paper we consider a two-level hierarchy, thus the value will first be allocated at the upper level, and then players at this level will allocate the designated values among the members of these groups at the lower level. Some potential areas of application and associated problems are discussed in the following sources. Two-level resource allocation models for LTE networks are studied in [6,9] as a bankruptcy game. A two-level cooperative model involving service providers is examined in [8] and includes coalition formation and coalition optimization steps for a revenue allocation problem. In [10,11] the authors propose a two-step cost allocation method for the transmission system among generators and loads in case of non-atomic players.

The structure of the paper is as follows. The model of a two-level network game is presented in Sect. 2. Section 3 is devoted to the allocation issues in the cooperative version of the game. There we show how to allocated a cooperative outcome in two steps and provide relations between several definitions of the characteristic function and the corresponding Shapley values. A special class of games so-called games with pairwise interactions is considered in Sect. 4. Conclusion is in Sect. 5.

2   The Model

Consider a situation in which a group of players make a decision in a constrained communication environment to achieve certain goals. We assume a two-level communication structure of players. The upper level of communication is represented by players from a given finite set A, $|A| \geqslant 2$, who are connected in a network g^1, which is a collection of undirected pairs $(i, j) \in A \times A$ called links. We exclude loops by supposing that $(i, i) \notin g^1$. At the lower level, there are $|A| = n$ finite groups of players $A_1, \ldots, A_n$ such that $A_i \cap A_j = \varnothing$, $i \neq j$. We assume that each player $i \in A$ is associated with the set A_i. Further, players from set $B_i = \{i\} \cup A_i$ are connected in a network g^{2i}. Here network g^{2i} is a collection of undirected links (i, j), $j \in A_i$, i.e., in g^{2i} players from A_i are connected only with i. Thus we have a player set $N = A \cup A_1 \cup \cdots \cup A_n = B_1 \cup \cdots \cup B_n$ of $|A| + \sum_{i \in N} |A_i|$ players connected in a network $g = g^1 \cup g^{21} \cup \cdots \cup g^{2n}$. Denote the neighbors of a player $i \in N$ in network g by $N_i(g) = \{j : (i, j) \in g\}$. An example of the network g is demonstrated in Fig. 1. It is a network of seven groups of players: players (nodes) from set A are filled black; other players compose six groups $A_1, \ldots, A_6$.

In the network a player can remove (all or some) links if such links are not beneficial for her. Let $(d_i(g), u_i)$ denote a strategy of player $i \in N$ in the game. Here $d_i(g) = (d_{ij}(g))_{i \in N}$ whose components equal either 1 if i keeps links with corresponding players, or 0, otherwise. The second component $u_i \in U_i$ of player i's strategy reflects her action. In the game, players select strategies in two stages. First, players at the upper level (players from A) simultaneously announce their choices, and knowing them, players at the lower level (players from $N \setminus A$) simultaneously select theirs. Denote $(d(g), u) = ((d_i(g), u_i))_{i \in N}$

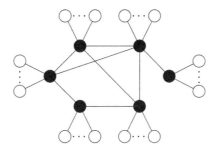

Fig. 1. A network of seven groups of players

and let $(d_S(g), u_S) = ((d_i(g), u_i))_{i \in S}$ for some subset $S \subset N$ called a *coalition*. Suppose that a player $i \in A$ selects $(d_i(g), u_i)$ and then a player $j \in A_i$ selects her strategy $(d_j(g), u_j)$ having the information about the choice of i. We note that profile $d(g)$ may change the network and thus we denote it by g^d.

Players are rewarded by payoffs depending upon their own strategies and the strategies of their neighbors in the given network g:

$$J_i(d(g), u) = h_i(u_i, u_{N_i(g^d)}) = \begin{cases} h_i(u_i, u_{N_i(g^{1,d})}, u_{N_i(g^{2i,d})}), & i \in A, \\ h_i(u_i, u_{N_i(g^{2j,d})}), & i \in A_j, \ j \in A. \end{cases}$$

This definition holds for all possible subnetworks g^d which can be realized when the strategy profile $(d(g), u)$ is played. The definition naturally comes from an option for players to remove some links from network g, and hence the payoff to the player is well-defined given any set of her neighbors (including the empty set). In the paper, we adopt the assumption from [13,14] that functions h_i, $i \in N$, satisfy the following property: for any player $i \in N$, any strategy $(d_i(g), u_i)$ and any coalition $S \subset N$, it holds true that $h_i(u_i, u_{N_i(g^d) \cap S'}) \leqslant h_i(u_i, u_{N_i(g^d) \cap S})$ for any $S' \subset S$. This property motivates players to keep all the links in the network. It also implies one useful result which will be discussed later in this paper.

3 Cooperation

Now suppose that in this two-stage decision process players coordinate their actions; the players seek to maximize the sum $\sum_{i \in N} J_i(d(g), u)$ followed by its allocation among them. Let

$$(\bar{d}(g), \bar{u}) = \arg \max_{(d(g), u)} \sum_{i \in N} J_i(d_i(g), u_i)$$

$$= \arg \max_{(d(g), u)} \sum_{i \in A} \left(h_i(u_i, u_{N_i(g^{1,d})}, u_{N_i(g^{2i,d})}) + \sum_{j \in A_i} h_j(u_j, u_{N_j(g^{2i,d})}) \right),$$

and thus players should allocate $\sum_{i \in N} J_i(\bar{d}(g), \bar{u})$. Using the property of functions h_i, $i \in N$, we immediately conclude that under cooperation players do not have to revise the network by removing links, and hence the expression for the sum to be allocated can be simplified:

$$\sum_{i \in N} J_i(\bar{d}(g), \bar{u}) = \sum_{i \in A} \left(h_i(\bar{u}_i, \bar{u}_{N_i(g^1)}, \bar{u}_{A_i}) + \sum_{j \in A_i} h_j(\bar{u}_j, \bar{u}_i) \right) \tag{1}$$

We consider transferable payoffs. Then to allocate (1), we need to determine corresponding TU games and find their solutions. We will do it in two steps: first we allocate (1) among the subgroups $B_1, \ldots, B_n$ and then the assigned values to subgroups B_i, $i = 1, \ldots, n$, will be allocated among their members. Following this idea, we first define a TU game (B, v_1) for the upper level, where $B = \{B_1, \ldots, B_n\}$ denote the set of players-groups and v_1 is a *characteristic function* assigning to any coalition $S \subseteq B$ its worth. In [14], the authors propose the definition of a characteristic function for a network game with an exogenously given network. The idea of that definition is very simple and intuitive: the worth of a coalition is only determined by the strength of its members as if the members of the complement would have no impact on the coalition. This definition also goes exactly in line with the approach of von Neumann and Morgenstern [15] defining the worth of a coalition as the maxmin value of a corresponding zero-sum game between this coalition and its complement. It has been shown that under the assumptions on payoff functions, the maxmin optimization problem is reduced to a maximization problem [14]. Hence adopting this approach to the model under consideration, we come to the following:

$$v_1(S) = \begin{cases} \sum\limits_{i \in N} J_i(\bar{d}(g), \bar{u}), & S = B, \\ \max\limits_{u_S} \sum\limits_{i \in A \cap S} \left(h_i(u_i, u_{N_i(g^1) \cap S}, u_{A_i}) + \sum\limits_{j \in A_i} h_j(u_j, u_i) \right), & S \subset B, \\ 0, & S = \varnothing. \end{cases}$$

From the above, we observe that the worth of the coalition of all players is just their largest total payoff, whereas to find the worth of a smaller coalition, we take into account only the strategies of its members. Indeed, according to the approach of von Neumann and Morgenstern, the coalition will not use an existing link with its complement if it is beneficial to it (as the complement will remove the link itself). Finally, the empty coalition gains nothing.

An imputation $\xi[v_1]$ in TU game (B, v_1) is a profile from $\mathbb{R}^n$ (as we have n players-groups) such that $\sum_{B_i \in B} \xi_{B_i}[v_1] = v_1(B)$ and $\xi_{B_i}[v_1] \geqslant v_1(B_i)$ for each $B_i \in B$. The set of all imputations called the *imputation set* will be denoted by $\mathcal{I}[v_1]$. A *cooperative solution* to the TU game (B, v_1) is a map assigning a subset $\mathcal{M}[v_1] \subseteq \mathcal{I}[v_1]$ to the game. Thus at the upper level, the value $v_1(B)$ is allocated among the players-groups by an imputation $\xi[v_1] \in \mathcal{M}[v_1]$. Under such the allocation a group $B_i \in B$ receives $\xi_{B_i}[v_1]$. Next, this value is allocated among the members of this group. For each B_i this is done with the use of a TU game (B_i, v_{2i}) in which the characteristic function is given by

$$v_{2i}(S) = \begin{cases} \xi_{B_i}[v_1], & S = B_i, \\[2mm] \max\limits_{u_S} \left(h_i(u_i, u_{N_i(g^{2i})}) + \sum\limits_{j \in A_i \cap S} h_j(u_j, u_i) \right), & S \subset B_i,\ i \in S, \\[3mm] \max\limits_{u_S} \sum\limits_{j \in S} h_j(u_j), & S \subset B_i,\ i \notin S, \\[2mm] 0, & S = \varnothing. \end{cases}$$

In a similar way, one can define the imputation set $\mathcal{I}[v_{2i}]$. To allocate $\xi_{B_i}[v_1]$ among the members of coalition $B_i = \{i\} \cup A_i$ we use the same cooperative solution $\mathcal{M}[v_{2i}] \subseteq \mathcal{I}[v_{2i}]$ which differs only in the characteristic function (v_{2i} replaces v_1). Thus selecting an imputation $\xi[v_{2i}] \in \mathcal{M}[v_{2i}]$ we solve the problem of allocating $\xi_{B_i}[v_1]$, $B_i \in B$.

When the cooperative solution $\mathcal{M}$ is the Shapley value and therefore consists of a single imputation, the corresponding allocations at both levels will have the following form:

$$\mathrm{Sh}_{B_i}[v_1] = \sum_{S \subseteq B,\, B_i \in S} \frac{(|B| - |S|)!(|S| - 1)!}{|B|!} \left(v_1(S) - v_1(S \setminus B_i) \right),$$

for any player-group $B_i \in B$ at the upper level, and

$$\mathrm{Sh}_j[v_{2i}] = \sum_{S \subseteq B_i,\, j \in S} \frac{(|B_i| - |S|)!(|S| - 1)!}{|B_i|!} \left(v_{2i}(S) - v_{2i}(S \setminus \{j\}) \right)$$

$$= \sum_{S \subseteq B_i,\, j \in S} \frac{(|A_i| - |S| + 1)!(|S| - 1)!}{(|A_i| + 1)!} \left(v_{2i}(S) - v_{2i}(S \setminus \{j\}) \right),$$

for any player j at the lower level from group $B_i = \{i\} \cup A_i \in B$, and player i at the upper level.

Remark 1. It worth mentioning that characteristic functions v_1 and v_{2i} are consistent in terms of their definitions except for coalition B_i. Indeed, B_i can guarantee for itself the value of

$$v_1(B_i) = \max_{u_{B_i}} \left(h_i(u_i, u_{A_i}) + \sum_{j \in A_i} h_j(u_j, u_i) \right)$$

since $u_{N_i(g^1) \cap B_i} = \varnothing$ and player i becomes disconnected from other players from A in g^1. However, selecting an imputation $\xi[v_1]$ at the upper level, coalition B_i will gain ξ_{B_i} according to it. Noting that the imputation satisfies individual rationality, it follows that $\xi_{B_i}[v_1] \geqslant v_1(B_i)$. Thus if we proceed to the lower level for allocating $\xi_{B_i}[v_1] \equiv v_{2i}(B_i)$, we get that $v_{2i}(B_i) - v_1(B_i) \geqslant 0$. This makes some inconsistency in the definition of the characteristic functions. For all other coalitions, functions v_1 and v_{2i} are associated with maximization problems of the same type and hence are consistent in their definitions. One can deal with the inconsistent but not crucial definition of the characteristic functions in two ways. In the first scenario, group B_i allocates what it has been allocated at the upper level, $\xi_{B_i}[v_1]$. For this reason, we do not define $v_{2i}(B_i)$ as the solution of a corresponding maximization problem, but just set $v_{2i}(B_i) \equiv \xi_{B_i}[v_1]$ and define v_{2i} only for subsets of B_i. This scenario has been used above: v_{2i} has been defined according to the mentioned procedure. In the second scenario, we can make a small change in the definition of the characteristic function at the lower level: instead of values $v_{2i}(S)$, $S \subseteq B_i$, we may introduce the adjusted values: the difference $\xi_{B_i}[v_1] - v_1(B_i)$ (a surplus for coalition B_i) can be allocated only to player i, and then the value $v_1(B_i)$, which this coalition can guarantee for itself, is allocated among the members. In this case $v_{2i}(S)$ remains unchanged for coalitions not containing player i, but for a coalition containing i, its worth is $v_{2i}(S) + \xi_{B_i}[v_1] - v_1(B_i)$, i.e.,

$$\tilde{v}_{2i}(S) = \begin{cases} \xi_{B_i}[v_1] \equiv v_1(S) + (\xi_{B_i}[v_1] - v_1(B_i)), & S = B_i, \\ v_{2i}(S) + (\xi_{B_i}[v_1] - v_1(B_i)), & S \subset B_i, \ i \in S, \\ v_{2i}(S), & S \subset B_i, \ i \notin S, \\ 0, & S = \varnothing. \end{cases}$$

Alternatively, the v_{2i} can be determined differently:

$$\tilde{\tilde{v}}_{2i}(S) = \begin{cases} \xi_{B_i}[v_1] \equiv v_1(S) + (\xi_{B_i}[v_1] - v_1(B_i))|S|/|B_i|, & S = B_i, \\ v_{2i}(S) + (\xi_{B_i}[v_1] - v_1(B_i))|S|/|B_i|, & S \subset B_i, \\ 0, & S = \varnothing. \end{cases}$$

for any $S \subseteq B_i$ by letting $v_{2i}(B_i) = v_1(B_i)$ and supposing that each player from B_i gets an additional and the same surplus of $(\xi_{B_i}[v_1] - v_1(B_i))/|B_i|$ from the allocation of $v_1(B_i)$. Here for all coalitions S, the definition of v_1 and v_{2i} becomes consistent (recall that $v_1(S)$ and $v_{2i}(S)$ are defined in the same manner, and the subscripts refer only to the level of the hierarchy).

We now demonstrate the relationship between the Shapley values for the above characteristic functions. Given the network g, let for any coalition $S \subseteq N$

$$v(S) = \max_{u_i, i \in S} \sum_{i \in S} h_i(u_i, u_{N_i(g) \cap S}).$$

For $T \subset N$ we define a restriction $v|_T$ of v: $v|_T(S) \equiv v(S)$ for all $S \subseteq T$. It is clear that $v_1(B) = v(N)$, $v_1(B_i) = v(B_i) = v|_{B_i}(B_i)$ and $v_{2i}(S) = v(S)$ for any $S \subset B_i$ and $i = 1, \ldots, n$.

Proposition 1. *For any group B_i and any player $j \in B_i$, the following relations hold true:*

$$\mathrm{Sh}_j[v_{2i}] = \mathrm{Sh}_j[v|_{B_i}] + (\mathrm{Sh}_{B_i}[v_1] - v_1(B_i))/|B_i|,$$

$$\mathrm{Sh}_j[\tilde{v}_{2i}] = \begin{cases} \mathrm{Sh}_j[v|_{B_i}] + (\mathrm{Sh}_{B_i}[v_1] - v_1(B_i)), & j = i, \\ \mathrm{Sh}_j[v|_{B_i}], & j \neq i, \end{cases}$$

$$\mathrm{Sh}_j[\tilde{\tilde{v}}_{2i}] = \mathrm{Sh}_j[v_{2i}].$$

Proof. To prove the first equality, we note:

$$\text{Sh}_j[v_{2i}] = \sum_{S \subseteq B_i, j \in S} \frac{(|B_i| - |S|)!(|S| - 1)!}{|B_i|!} (v_{2i}(S) - v_{2i}(S \setminus \{j\}))$$

$$= \frac{v_{2i}(B_i) - v_{2i}(B_i \setminus \{j\})}{|B_i|}$$

$$+ \sum_{S \subset B_i, j \in S} \frac{(|B_i| - |S|)!(|S| - 1)!}{|B_i|!} (v_{2i}(S) - v_{2i}(S \setminus \{j\}))$$

$$= \frac{v_{2i}(B_i) - v_1(B_i) + v_1(B_i) - v_{2i}(B_i \setminus \{j\})}{|B_i|}$$

$$+ \sum_{S \subset B_i, j \in S} \frac{(|B_i| - |S|)!(|S| - 1)!}{|B_i|!} (v|_{B_i}(S) - v|_{B_i}(S \setminus \{j\}))$$

$$= \frac{v_{2i}(B_i) - v_1(B_i) + v|_{B_i}(B_i) - v|_{B_i}(B_i \setminus \{j\})}{|B_i|}$$

$$+ \sum_{S \subset B_i, j \in S} \frac{(|B_i| - |S|)!(|S| - 1)!}{|B_i|!} (v|_{B_i}(S) - v|_{B_i}(S \setminus \{j\}))$$

$$= \frac{\text{Sh}_j[v_1] - v_1(B_i)}{|B_i|} + \text{Sh}_j[v|_{B_i}],$$

where $v_{2i}(B_i) = \text{Sh}_j[v_1]$.

Now we show the fulfillment of the second equality. Since only coalitions $S \subseteq B_i, S \ni i$, gain a constant value of $\text{Sh}_{B_i}[v_1] - v_1(B_i)$, then by the properties of the Shapley value player i gets this value as her additional payoff. Further, as $v_1(B_i) = v|_{B_i}(B_i)$ and $v_{2i}(S) = v|_{B_i}(S)$ for all $S \subset B_i$, we get that $\text{Sh}_j[\tilde{v}_{2i}] = \text{Sh}_j[v|_{B_i}]$ for $j \neq i$ and $\text{Sh}_i[\tilde{v}_{2i}] = \text{Sh}_i[v|_{B_i}] + (\text{Sh}_{B_i}[v_1] - v_1(B_i))$.

Finally, the third equality also follows by the properties of the Shapley value. Since each coalition $S \subseteq B_i$ gains a value of $(\text{Sh}_{B_i}[v_1] - v_1(B_i))|S|/|B_i|$, then each player just gets an additional value of $(\text{Sh}_{B_i}[v_1] - v_1(B_i))/|B_i|$. Further, as $v_1(B_i) = v|_{B_i}(B_i)$ and $v_{2i}(S) = v|_{B_i}(S)$ for all $S \subset B_i$, we get that $\text{Sh}_j[\tilde{\tilde{v}}_{2i}] = \text{Sh}_j[v|_{B_i}] + (\text{Sh}_{B_i}[v_1] - v_1(B_i))/|B_i|$ and therefore, $\text{Sh}_j[\tilde{\tilde{v}}_{2i}] = \text{Sh}_j[v_{2i}]$. This concludes the proof. □

The proposed two-level allocation procedure differs from the standard one-level scheme used in the classical cooperative game theory. We illustrate this difference for the Shapley value. To compute the one-level Shapley value, one has to use all subsets of the player set N, i.e. $2^{|N|}$ sets. However for the two-level Shapley value, we first use only subsets of N consisting of all possible unions of groups $B_1, \ldots, B_n$, and then at the lower level we use all subsets of the groups, i.e. get only $2^{|B|} + 2^{|A_1|} + \ldots + 2^{|A_n|}$ subcoalitions ($|B| + |A_1| + \ldots + |A_n| = |N|$). Thus in the two-level allocation scheme we do not list all subsets of N. To point out the difference between the classical (one-level) allocation procedure and the two-level procedure, we consider the next example.

Example 1. Consider a five-person game. Let the player set $N = \{1, 2, 3, 4, 5\}$ be decomposed into three groups: $N = A \cup A_1 \cup A_2$ where $A = \{1, 4\}$ is the set of players at the upper level, $A_1 = \{2, 3\}$ is the set of players at the lower level subordinated to player 1, and $A_2 = \{5\}$ is the set of players at the lower level subordinated to player 4. The network g is demonstrated in Fig. 2. Let $B_1 = \{1\} \cup A_1$, $B_2 = \{4\} \cup A_2$.

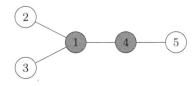

Fig. 2. A network g of three groups of players for Example 1

Let $u_i \in U_i = [0, \infty)$ be an action for player $i \in N$, and this player be rewarded with her payoff function $h_i(u_i, u_{N_i(g)}) = \ln(1 + u_i + \sum_{j \in N_i(g)} u_j) - c u_i$, for $c > 0$. This is an example of a logarithmic utility function for a model of public goods provision [2]. Here we suppose that $c = 0.1$.

One can show that under cooperation, the total players payoff equals $v(N) = v(\{B_1, B_2\}) = 12.810$ (we round off all numbers to third decimal place). For a two-level model, we first have to allocate $v(\{B_1, B_2\})$ at the upper level between groups B_1 and B_2. To do this find $v_1(B_1) = 7.303$ and $v_1(B_2) = 4.091$ and then get the Shapley value $\text{Sh}[v_1] = (\text{Sh}_{B_1}[v_1], \text{Sh}_{B_2}[v_1])$ with $\text{Sh}_{B_1}[v_1] = 8.011$ and $\text{Sh}_{B_2}[v_1] = 4.799$. Next the corresponding values are allocated at the lower level. For this reason for group B_1 we find $v_{21}(B_1) = \text{Sh}_{B_1}[v_1]$, $v_{21}(\{1, 2\}) = v_{21}(\{1, 3\}) = 4.091$, $v_{21}(\{2, 3\}) = 2.805$, $v_{21}(\{1\}) = v_{21}(\{2\}) = v_{21}(\{3\}) = 1.403$ and therefore the Shapley value $\text{Sh}[v_{21}] = (\text{Sh}_1[v_{21}], \text{Sh}_2[v_{21}], \text{Sh}_3[v_{21}])$ consists of the components: $\text{Sh}_1[v_{21}] = 3.099$, $\text{Sh}_2[v_{21}] = \text{Sh}_3[v_{21}] = 2.456$. For group B_2 we find $v_{22}(B_2) = \text{Sh}_{B_2}[v_1]$, $v_{22}(\{4\}) = v_{22}(\{5\}) = 1.403$ and therefore the Shapley value $\text{Sh}[v_{22}] = (\text{Sh}_4[v_{22}], \text{Sh}_5[v_{22}])$ consists of the components: $\text{Sh}_4[v_{22}] = \text{Sh}_5[v_{22}] = 2.400$ (the components do not sum up to 4.799 due to rounding).

If the players had allocated the value of $v_1(N) = v_1(\{B_1, B_2\}) = 12.810$ only in one step under the classical allocation procedure, we would have had the following one-level characteristic function v: $v(\{N\}) = 12.810$, $v(\{1, 2, 3, 4\}) = 10.856$, $v(\{1, 2, 3, 5\}) = v(\{1, 2, 4, 5\}) = 9.519$, $v(\{2, 3, 4, 5\}) = 6.897$, $v(\{1, 2, 3\}) = v(\{1, 2, 4\}) = v(\{1, 3, 4\}) = v(\{1, 4, 5\}) = 7.304$, $v(\{1, 2, 5\}) = v(\{1, 3, 5\}) = v(\{2, 4, 5\}) = v(\{3, 4, 5\}) = 5.494$, $v(\{2, 3, 4\}) = v(\{2, 3, 5\}) = 4.208$, $v(\{1, 2\}) = v(\{1, 3\}) = v(\{1, 4\}) = v(\{4, 5\}) = 4.091$, $v(\{1, 5\}) = v(\{2, 3\}) = v(\{2, 4\}) = v(\{2, 5\}) = v(\{3, 4\}) = v(\{3, 5\}) = 2.805$, $v(\{1\}) = v(\{2\}) = v(\{3\}) = v(\{4\}) = v(\{5\}) = 1.403$, and the one-level Shapley value $\text{Sh}[v] = (\text{Sh}_1[v], \text{Sh}_2[v], \text{Sh}_3[v], \text{Sh}_4[v], \text{Sh}_5[v])$ whose components equal $\text{Sh}_1[v] = 3.633$, $\text{Sh}_2[v] = \text{Sh}_3[v] = 2.247$, $\text{Sh}_4[v] = 2.812$, $\text{Sh}_5[v] = 1.869$ (the components do not sum up to 12.810 due to rounding).

Comparing the two values we observe that players 2, 3 and 5 (lower-level players) benefit from the two-level allocation procedure whereas players 1 and 4 (upper-level players) benefit from the classical one-level scheme. Moreover, following the two-level scheme players 4 and 5 (players of different levels) are rewarded equally which is not the case when the one-level allocation procedure is applied. The example demonstrates that there should be a trade-off between the schemes as players benefit unequally from them. We conclude this example by noting that $\text{Sh}_1[v|_{B_1}] = 2.863$, $\text{Sh}_2[v|_{B_1}] = \text{Sh}_3[v|_{B_1}] = 2.220$, $\text{Sh}_4[v|_{B_2}] = \text{Sh}_5[v|_{B_2}] = 2.046$. And finally, $\text{Sh}_1[\tilde{v}_{21}] = 3.571$, $\text{Sh}_2[\tilde{v}_{21}] = \text{Sh}_3[\tilde{v}_{21}] = 2.220$, $\text{Sh}_4[\tilde{v}_{22}] = 2.753$, $\text{Sh}_5[\tilde{v}_{22}] = 2.046$.

4   A Case of Pairwise Interactions

This section deals with a case of pairwise interactions introduced in [4] and then developed in [12] for cooperative network games. Pairwise interaction means that a player gains from each of her neighbors by choosing actions not necessarily the same for each of the neighbors. An interaction between two connected players in the network is generally represented by a bimatrix game in which each of the players has a finite number of actions. For example, a network game with a 2×2 coordination game played between neighbors was studied in [7] and later in [5]. A coordination stag-hunt game with repeated rounds was considered in [3]. In [16], the authors developed a network game model based on prisoner's dilemma.

Specifically, let $i \in N$, $j \in N_i(g)$, then player i plays with her neighbor j a bimatrix game with payoff matrices $A_{ij} = \{a_{p\ell}^{ij}\}$ and $B_{ij} = \{b_{p\ell}^{ij}\}$ for players i and j, respectively. Given a profile of players' strategies, the player's payoff represents the sum of her payoffs in all bimatrix games played with her neighbors in the network. Let

$$m_{ij} = \begin{cases} \max_{p,\ell}(a_{p\ell}^{ij} + b_{p\ell}^{ji}), & \text{if } i \text{ and } j \text{ are neighbors,} \\ 0, & \text{otherwise.} \end{cases}$$

Then characteristic functions v_1 and v_{2i}, $i \in A$ will have the following form:

$$v_1(S) = \begin{cases} \dfrac{1}{2} \sum_{i \in A} \left(\sum_{j \in N_i(g^1)} m_{ij} + 2 \sum_{j \in A_i} m_{ij} \right), & S = B, \\ \dfrac{1}{2} \sum_{i \in A \cap S} \left(\sum_{j \in N_i(g^1) \cap S} m_{ij} + 2 \sum_{j \in A_i} m_{ij} \right), & S \subset B, \\ 0, & S = \varnothing, \end{cases}$$

and for any $B_i \in B$,

$$v_{2i}(S) = \begin{cases} \xi_{B_i}[v_1], & S = B_i, \\ \sum_{j \in A_i \cap S} m_{ij}, & S \subset B_i, \ i \in S, \\ 0, & \text{otherwise.} \end{cases}$$

By the definition of v_1, we get $v_1(B_i) = \sum_{j \in A_i} m_{ij}$. Consider the characteristic function

$$v_{2i}^{\text{PI}}(S) = \begin{cases} \sum\limits_{j \in A_i \cap S} m_{ij}, & S \subseteq B_i, \ i \in S, \\ 0, & \text{otherwise.} \end{cases}$$

At the lower level, network g^{2i} is a star-network whose hub is player i. For a star network, [12] provides the formula for the Shapley value $\text{Sh}[v_{2i}^{\text{PI}}]$. Adopting this formula for $v_{2i}^{\text{PI}}(S)$, we obtain:

$$\text{Sh}_j[v_{2i}^{\text{PI}}] = \begin{cases} \dfrac{1}{2} \sum\limits_{k \in A_i} m_{ik}, & j = i, \\ \dfrac{m_{ij}}{2}, & j \neq i. \end{cases}$$

Corollary 1. *In the case of pairwise interactions for any group B_i and any player $j \in B_i$, the following relations hold true:*

$$\text{Sh}_j[v_{2i}] = \text{Sh}_j[v_{2i}^{\text{PI}}] + (\text{Sh}_{B_i}[v_1] - v_1(B_i))/|B_i|,$$

$$\text{Sh}_j[\tilde{v}_{2i}] = \begin{cases} \text{Sh}_j[v_{2i}^{\text{PI}}] + (\text{Sh}_{B_i}[v_1] - v_1(B_i)), & j = i, \\ \text{Sh}_j[v_{2i}^{\text{PI}}], & j \neq i, \end{cases}$$

$$\text{Sh}_j[\tilde{\tilde{v}}_{2i}] = \text{Sh}_j[v_{2i}].$$

Proof. This statement directly follows from Proposition 1 by noting that $v_{2i}^{\text{PI}} = v|_{B_i}$, $i = 1, \ldots, n$.

5   Conclusion

In the paper, we have proposed an allocation procedure for a two-level communication structure. At the upper level, we allocate the total payoff among the groups of players, and then the designated values are allocated within these groups. This approach can find its application in hierarchical structures or organizations. In the paper, we illustrated the results for the Shapley value chosen as an cooperative solution (allocation rule) at both levels. Definitely, the model is flexible with respect to the selection of cooperative solutions and is not limited to the Shapley value only. Moreover, in the model we can adopt two different allocation rules at the upper and lower levels of communication (even solutions from the non-transferable utility theory). In some instances, the solutions for games with non-transferable utility, for example, the Nash bargaining solution, could fit the model more precisely [1], in particular, when at the upper level groups have payoffs of different types. Once the total payoff has been allocated, players can use a cooperative solution for transferable payoffs within the group as its members can have payoffs of the same type. In this case, one should use proper allocations at both levels of communication. The model may be extended to the case of multiple levels of hierarchy, and we hope that the proposed in the paper technique can be adopted to it. This is left for future research.

Acknowledgements. The authors thank reviewers for valuable comments.

References

1. Avrachenkov, K., Elias, J., Martignon, F., Neglia, G., Petrosyan, L.: Cooperative network design: a nash bargaining solution approach. Comput. Netw. **83**(4), 265–279 (2015)
2. Bramoullé, Y., Kranton, R.: Public goods in networks. J. Econ. Theory **135**(1), 478–494 (2007)
3. Corbae, D., Duffy, J.: Experiments with network formation. Games Econ. Behav. **64**, 81–120 (2008)
4. Dyer, M., Mohanaraj, V.: Pairwise-interaction games. In: Aceto, L., Henzinger, M., Sgall, J. (eds.) ICALP 2011. LNCS, vol. 6755, pp. 159–170. Springer, Heidelberg (2011). https://doi.org/10.1007/978-3-642-22006-7_14
5. Goyal, S., Vega-Redondo, F.: Network formation and social coordination. Games Econ. Behav. **50**, 178–207 (2005)
6. Iturralde, M., Wei, A., Ali-Yahiya, T., et al.: Resource allocation for real time services in LTE networks: resource allocation using cooperative game theory and virtual token mechanism. Wirel. Pers. Commun. **72**, 1415–1435 (2013)
7. Jackson, M., Watts, A.: On the formation of interaction networks in social coordination games. Games Econ. Behav. **41**(2), 265–291 (2002)
8. Liao, J., Cui, Z., Wang, J., et al.: A coalitional game approach on improving interactions in multiple overlay environments. Comput. Netw. **87**, 1–15 (2015)
9. Madi, N.K.M., Hanapi, Z.B.M., Othman, M., et al.: Two-level QoS-aware frame-based downlink resources allocation for RT/NRT services fairness in LTE networks. Telecommun. Syst. **66**, 357–375 (2017)
10. Molina, Y.P., Prada, R.B., Saavedra, O.R.: Complex losses allocation to generators and loads based on circuit theory and Aumann-Shapley method. IEEE Trans. Power Syst. **25**(4), 1928–1936 (2010)
11. Molina, Y.P., Saavedra, O.R., Amarís, H.: Transmission network cost allocation based on circuit theory and the Aumann-Shapley method. IEEE Trans. Power Syst. **28**(4), 4568–4577 (2013)
12. Petrosyan, L.A., Bulgakova, M.A., Sedakov, A.A.: Time-consistent solutions for two-stage network games with pairwise interactions. Mob. Netw. Appl. (2018). https://doi.org/10.1007/s11036-018-1127-7
13. Petrosyan, L.A., Sedakov, A.A.: The subgame-consistent shapley value for dynamic network games with shock. Dyn. Games Appl. **6**(4), 520–537 (2016)
14. Petrosyan, L.A., Sedakov, A.A., Bochkarev, A.O.: Two-stage network games. Autom. Remote Control **77**(10), 1855–1866 (2016)
15. Von Neumann, J., Morgenstern, O.: Theory of Games and Economic Behavior. Princeton University Press, Princeton (1944)
16. Xie, F., Cui, W., Lin, J.: Prisoners dilemma game on adaptive networks under limited foresight. Complexity **18**, 38–47 (2013)

Using Bankruptcy Rules to Allocate CO2 Emission Permits

Raja Trabelsi[1,2,3]($\boxtimes$), Stefano Moretti[2], and Saoussen Krichen[3]

[1] LAMSADE and LARODEC, Paris, France
[2] Universite Paris-Dauphine, PSL Research University, CNRS, LAMSADE,
Place du Maréchal de Lattre de Tassigny, 75775 Paris Cedex 16, France
`raja.trabelsi@dauphine.eu, stefano.moretti@dauphine.fr`
[3] LARODEC, Institut Superieur de Gestion de Tunis,
University of Tunis,
Tunis, Tunisia
`Saoussen.krichen@isg.rnu.tn`

Abstract. The global growth of technologies and production affects the climate through emissions of greenhouse gases. The total amount of countries' demands of CO2 emissions permits is higher than what the planet can sustain. This situation can be considered as a bankruptcy problem, where the sum of players' claims exceeds the endowment of the resource. In this paper, we use an approach based on bankruptcy solutions (in particular, on the Weighted Constrained Equal Awards rule) in order to propose a more efficient and fair allocation protocol for sharing CO2 emissions permits among the EU-28 countries.

Keywords: Bankruptcy situations ·
Weighted Constrained Equal Awards · CO2 emissions ·
Cooperative games

1 Introduction

Establishing international agreements aimed to reduce greenhouse gas emissions is a current problem [20], due to the fact that greenhouse gases present a big threat to the planet. Negotiation models [18], in particular, have been widely studied with the goal to find an agreement among all countries for reducing their carbon-emission rate. However, no country can influence the climate change system independently of the others, and everyone would benefit from the reduction on global warming, but no one wants to bear the cost of emission reductions. One of the main issues in this research stream is the problem of finding fair and efficient protocols to allocate the permits for carbon dioxide (CO2) emission. In the related literature, the methods used to allocate CO2 emission permits can be classified into four groups [21]: protocols based on global indicators, optimization techniques, game theoretic models and hybrid approaches. The most common allocation protocols are based on indicators, where permits are assigned based

© ICST Institute for Computer Sciences, Social Informatics and Telecommunications Engineering 2019
Published by Springer Nature Switzerland AG 2019. All Rights Reserved
K. Avrachenkov et al. (Eds.): GameNets 2019, LNICST 277, pp. 82–92, 2019.
https://doi.org/10.1007/978-3-030-16989-3_6

on one or more global indicators for countries [14] like, for instance, population [9], energy [15], Gross Domestic Product (GDP) [17], or emission intensity [15], etc. Among the optimization methods, some authors have proposed the Data Envelopment Analysis based on linear programming models [5,7]. In the game theoretic literature related to non-cooperative games, a CO2 emission permits allocation is seen as an outcome that is obtained at the equilibrium of a strategic game [2]. In the domain of cooperative games, instead, various allocation rules have been applied, like, for instance, the Shapley value [4]. The hybrid approach is a mixture of different methods, considering, among others, multi-stage regimes or multi-sector convergence approaches [1]. In this paper, we analyse the allocation problem of CO2 emission permits as a conflicting claims problem or *bankruptcy problem* [13], where countries claim a scarce resources or estate (the maximum amount of CO2 emissions) and there is not enough resource to satisfy the aggregate claim [6,8] (see also [10,11,19] for other applications of bankruptcy problems to natural resource sharing). In this paper, the computation of the total amount of CO2 emission is based on the EU emission quantity in 1990. More precisely, we consider a bankruptcy problem where the players are the EU-28 states, claims are the quantity of CO2 emitted in each year from 2010 to 2014, and the resource to be shared corresponds to the 78% of the total production of CO2 in EU in 1990, i.e. the quantity that countries should have produced in 2010 according to the Kyoto Protocol[1]. Our approach is based on a weighted version of the Constrained Equal Awards (CEA) rule [12] for bankruptcy situations, where we consider two parameters to allocate the CO2 emission quantity for each state: the quantity of CO2 emitted by each country in 1990 (claim) and the GDP of each country (weight). To be more specific, we consider a bankruptcy situation where every state claims a quantity of CO2 and it wants to "use" this emission to produce a certain amount of GDP. Nonetheless, a country can not obtain more than $\lambda \times GDP$ of CO2 emission permit, where λ is a fixed coefficient defined to guarantee the budget balance of the allocation, taking into consideration the relation among demands and GDP of all countries. In our application, the GDP is assumed to reflect the total overall economic activity, therefore countries with low $CO2/GDP$ ratio are considered more efficient because of a lower CO2 emission per unit of production of GDP. We compare the results provided by the weighted CEA rule with those of the classical CEA rule, that has been already proposed as an allocation rule for greenhouse gas emission permits [6]. According to the CEA rule, countries with small demands are completely satisfied, while highest demands are only partially satisfied. Instead, using the weighted CEA rule, the allocation changes in favour of countries with more efficient production technologies (low CO2/GDP ratio). Incentives to the transfer of green production technologies from the most efficient countries to the less efficient ones are also studied using a (cooperative) game theoretic approach. The road-map of the paper is as follows. We start in the next section with some preliminary definitions. Then, in Sect. 3, we introduce and compare the allocations of CO2 permits provided by the CEA rule and

---

[1] https://unfccc.int/process/the-kyoto-protocol.

the weighted CEA one over the EU-28 countries. Section 4 is devoted to a preliminary analysis of an associated cooperative game aimed at finding incentives to transfer technologies from the most efficient countries to the most polluting ones. Section 5 concludes.

2   Preliminary Notions and Notations

Let $N = \{1, \cdots, n\}$ be a finite set of agents. A bankruptcy situation [13,16] consists of a pair (E, c) where $c \in \mathbb{R}_+^N$ is a vector representing agent's claims and such that $c_i \geq 0$ for all $i \in N$, and $E \in \mathbb{R}$ is the estate to be divided among the agents and such that $0 < E < \sum_{i=1}^N c_i$. We denote as B^N the class of all bankruptcy situations with N as the set of agents. An *allocation rule* or *solution* for bankruptcy situations is a map $\phi : B^N \rightarrow \mathbb{R}_+^N$ that associates to every bankruptcy situation in B^N an allocation vector $x \in \mathbb{R}_+^N$ such that $\sum_{i=1}^N x_i = E$. So a solution for bankruptcy situations specifies vectors representing the amount of estate that each player should receive. Given a bankruptcy situation (E, c), the Constrained Equal Awards (CEA) rule allocates the estate E to agents according to the following formula:

$$CEA_i(E, c) = \min\{c_i, \lambda\} \tag{1}$$

where the parameter λ is such that $\sum_{i \in N} \min\{c_i, \lambda\} = E$. Given a bankruptcy situation (E, c) and a weight vector $a \in \mathbb{R}_+^N$, we call the triple (E, c, a) a *weighted bankruptcy situation*. We denote by WB^N the class of all weighted bankruptcy situations (E, c, a) with N as the set of agents. A solution for weighted bankruptcy situations is then a map $\psi : WB^N \rightarrow \mathbb{R}_+^N$. The solution called *Weighted* Constrained Equal Awards (WCEA) rule has been studied in [3,12] and it is defined as follows:

$$WCEA_i(E, c, a) = \min\{c_i, \hat{\lambda} a_i\} \tag{2}$$

where the parameter $\hat{\lambda} \in \mathbb{R}$ is such that

$$\sum_{i \in N} \min\{c_i, \hat{\lambda} a_i\} = E \tag{3}$$

A *Transferable Utility* (TU-) *game* with N as the set of players, is a pair (N, v), where $v : 2^N \rightarrow \mathbb{R}$ is the *characteristic function*, representing the *worth* or *profit* $v(S)$ of *coalition* $S \subseteq N$ (with $v(\emptyset) = 0$). Often, we identify a TU-game (N, v) with its characteristic function v. A TU-game (N, v) is *monotonic* if it holds that $v(S) \leq v(T)$ for all S and T such that $S \subseteq T \subseteq N$; it is *superadditive* if it holds that $v(S \cup T) \geq v(S) + v(T)$ for all $S, T \subseteq N$ such that $S \cap T = \emptyset$; it is *convex* or *supermodular* if it holds that $v(S \cup T) + v(S \cap T) \geq v(S) + v(T)$ for all $S, T \subseteq N$. Given a game v, an *imputation* is a vector $x \in \mathbb{R}^N$ such that $\sum_{i \in N} x_i = v(N)$ and $x_i \geq v(\{i\})$ for all $i \in N$. The *core* of a TU-game v is denoted by $C(v)$ and it is defined as the set of imputations defined as follows:

$$C(v) = \{x \in \mathbb{R}^N : \sum_{i \in N} x_i = v(N), \sum_{i \in S} x_i \geq v(S) \quad \forall S \subset N\}.$$

In general, a TU-game may have an empty core, and a TU-game with a non-empty core is said to be *balanced*. It is well known that convex games are balanced, but a balanced game need not be convex.

3   Bankruptcy Solutions for Emission Permits Allocation

According to the Kyoto Protocol, in 2010 the EU-28 countries should have reduced their cumulative CO2 by the 22% of the total amount of CO2 produced by the EU in 1990. Unfortunately, based on the CO2 data emissions provided by World Bank Open Data[2], this objective has not been achieved during any of the five years from 2010 to 2014. In this section, we want to analyse *ex-post* how CO2 emission permits would have been allocated among the EU-28 countries using the (W)CEA rule in order to meet the requirement indicated by the Kyoto protocol. To this aim, we model the problem of allocating CO2 permits during a year y, $y = 2010, \ldots, 2014$, as a bankruptcy situation (E, c^y), where the agents are the EU-28 countries, the estate E corresponds to the 78% of the cumulative amount of CO2 produced in EU in 1990 (equal to 2434.658 Gt [8]), and claim c_i^y, for each country i, represents the actual amount of CO2 emitted during year y, as indicated in World Bank Open Data.

Using the very famous CEA rule, countries with a small claim, receive all their demands (see Fig. 1). Instead, the highest claims are not completely satisfied. If a country's claim decreases due to a new emission reduction policy, it is likely that it receives an amount of emission permits equal to its claim. For instance, until 2012, Spain's claims is higher than 250.000 Gt, so Spain does not receive the full claimed amount of permits. However, when Spain's claim becomes less than 250.000 Gt (see Fig. 2), the allocated amount covers the full claim of emission permits for Spain. On the other hand, countries with high demands like Poland, France, Italy, UK and Germany, receive the same quantity, despite the fact that they have different claims. For example, Germany's claim is double than the one of France and it is three times the claim of Spain (See Fig. 3). In this paper we propose to use an alternative allocation rule, the weighted CEA, which is based on a weighted bankruptcy situation (E, c^y, a^y) for each year y, $y = 2010, \ldots, 2014$. As a weight vector, we set a_i^y equal to the GDP (Gros Domestic Product) of country i at year y. Country's GDP is an economic indicator reflecting the total value of "wealth production" within the country. Under this interpretation of a weighted bankruptcy situation, a country i claims the amount of CO2 emission permits c_i to produce the GDP a_i. However, country i will never obtain a share of CO2 emission permits larger than $\hat{\lambda} a_i$, where $\hat{\lambda}$ is the *maximum emission intensity per unit of GDP* calculated according to relation (2).

In the first part of this section, we used the CEA rule to allocate the CO2 emission permits and we had noticed that this method is beneficial for countries with low claims. Differently, the weighted CEA rule favours countries with low emission intensities $\lambda_i = \frac{c_i}{a_i}$. For instance, countries claiming more than

---

[2] https://data.worldbank.org/indicator/EN.ATM.CO2E.KT?view=map.

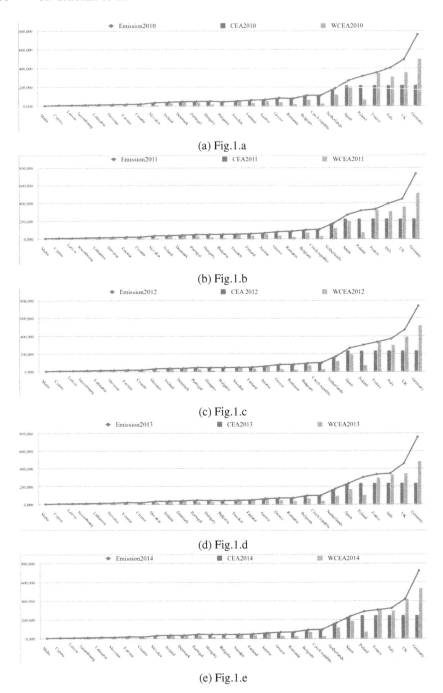

(a) Fig.1.a

(b) Fig.1.b

(c) Fig.1.c

(d) Fig.1.d

(e) Fig.1.e

Fig. 1. CEA and WCEA allocations CO2 emissions permits for 28-EU

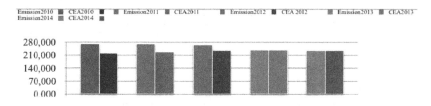

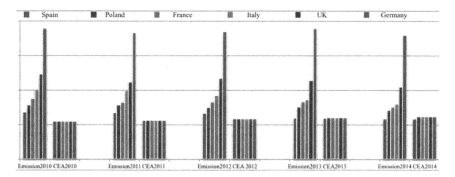

Fig. 2. Spain's allocation of CO2 emissions

Fig. 3. CO2 allocation for countries with the highest demand.

200.000 GT of CO2 emissions receive less than their claim using the CEA rule, but adopting the WCEA rule, they are better off. This is, for instance, the case of Germany, that in 2010 claims 758860 GT of CO2, and using the CEA rule obtains 220420 GT of CO2 emission permits, i.e. less than half of the claim. Instead, using the weighted CEA, Germany obtains 500647 GT of CO2 emission permits, i.e. about two-thirds of the claim. Differently, Poland has a high claim but also a large emission intensity λ_i. So, using the CEA rule, Poland receives a larger quantity of CO2 emission permits than using the weighted CEA rule (See Fig. 4). In the next section, we show how the sharing policy generated by the weighted CEA rule may boost the technological transfer from efficient countries (with a low emission intensity) to most polluting ones (with a high emission intensity) that can be compensated by profit transfers among cooperating countries.

4   Surplus and Technology Transfer

Based on the allocation provided by the WCEA solution, countries with a high ratio $\frac{c_i}{a_i}$ receive less than their claims, even if their claims are small, whereas countries with a low $\frac{c_i}{a_i}$ likely receive their full demands. As observed in the previous section, a low ratio $\frac{c_i}{a_i}$ reflects a good ability to emit low quantities of CO2 for unit of GDP. We argue that this ability is the consequence of a more efficient use of resources and a higher level of green technologies. In this section, we study the problem of implementing economic incentives to transfer technology

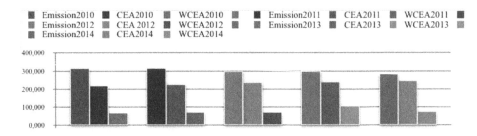

Fig. 4. Poland's allocation CO2 emissions permits

from the most efficient countries to the less efficient ones. To this purpose, we introduce a TU-game where the profit of each coalition of players (again, the EU-28 countries) is computed as the total profit obtained within the coalition using the highest technological level available for players in the coalition.

Definition 1. *Let (E, c, a) a weighted bankruptcy situation and let WCEA $(E, c, a) = (x_1, x_2, \ldots, x_n)$. For each $i \in N$, let $\lambda_i = \frac{c_i}{a_i}$ and let $\lambda(S) = \min_{i \in S: x_i - c_i = 0} \lambda_i$ for all $S \subseteq N$ (with the convention that $\lambda(S) = 0$ if $x_i - c_i \neq 0$ for all $i \in S$). The corresponding Technology-Transfer (TT-) game is defined as the TU-game $(N, \tilde{v})$ such that for all $S \subseteq N$*

$$\tilde{v}(S) = \begin{cases} \sum_{i \in S: c_i - x_i > 0} \left(\frac{x_i}{\lambda(S)} - \frac{x_i}{\lambda_i} \right) & \text{if } \lambda(S) \neq 0, \\ 0 & \text{otherwise.} \end{cases} \tag{4}$$

So the worth $\tilde{v}(S)$ of a coalition S represents the profit that a coalition $S \subseteq N$ can guarantee from adopting the best technology in use among the players in S, which corresponds to the best emission rate $\lambda(S) = \min_{i \in S: x_i - c_i = 0} \lambda_i$ among players in S who have received the claimed amount of emission permits. Precisely, the quantity of CO2 emission permits x_i allocated by the WCEA rule to each country $i \in S$ whose claim is not completely satisfied (i.e., $c_i - x_i > 0$) is used to produce a profit at the rate $\lambda(S)$, i.e. $\frac{x_i}{\lambda(S)} - \frac{x_i}{\lambda_i}$, and these profits are summed up over the countries in S. In other words, the "transfer" of technology within a coalition is only allowed from countries $i \in N$ with $\lambda_i \leq \hat{\lambda}$ to countries $j \in N$ with $\lambda_i > \hat{\lambda}$, where $\hat{\lambda}$ is the limit of emission intensity imposed by relation (2). Notice that $\frac{x_i}{\lambda(S)} - \frac{x_i}{\lambda_i} \geq 0$ for all $S \subseteq N$ such that $\lambda(S) \neq 0$ and every $i \in S$.

Proposition 1. *TT-games are monotonic and superadditive.*

Proof. We omit the straightforward proof that TT-games are monotonic. Consider a TT-game $(N, \tilde{v})$ corresponding to the allocation generated by the weighted CEA rule $WCEA(E, c, a) = (x_1, x_2, \ldots, x_n)$. Consider any coalitions $S, T \subseteq N$ with $S \cap T = \emptyset$. If $\tilde{v}(S) = 0$ or $\tilde{v}(T) = 0$ (or both), by monotonicity of $\tilde{v}$, it directly follows that $\tilde{v}(S) + \tilde{v}(T) \leq \tilde{v}(S \cup T)$. Otherwise, if $\tilde{v}(S) > 0$ and $\tilde{v}(T) > 0$, we have that

$$\tilde{v}(S) + \tilde{v}(T) \leq \sum_{i \in S \cup T : c_i - x_i > 0} \left(\frac{x_i}{\min\{\lambda(S), \lambda(T)\}} - \frac{x_i}{\lambda_i} \right) = \tilde{v}(S \cup T),$$

where the equality follows from the fact that $\min\{\lambda(S), \lambda(T)\} = \lambda(S \cup T)$. So, we have proved that $\tilde{v}(S)$ is superadditive.

The following example show that TT-games are not convex in general.

Example 1. Let $(N, \tilde{v})$ be the TT-game corresponding to the weighted bankruptcy situation (E, c, a) with $E = 14$ and the other parameters as shown in Table 1. It is easy to check by relations (2) and (3) that $\lambda^* = \frac{1}{2}$, yielding the weighted CEA allocation shown in the last column of Table 1. Let $S = \{2, 3, 5\}$ and $T = \{3, 4\}$. Then: $\tilde{v}(S) = \left(\frac{3}{1} - \frac{3}{2} \right) + \left(\frac{5}{\frac{1}{3}} - \frac{5}{\frac{2}{5}} \right) = 14$, $\tilde{v}(T) = \frac{5}{\frac{2}{5}} - \frac{5}{\frac{2}{5}} = \frac{23}{4}$, $\tilde{v}(S \cap T) = \tilde{v}(\{3\}) = 0$, $\tilde{v}(S \cup T) = \tilde{v}(\{2, 3, 4, 5\}) = \tilde{v}(\{2, 3, 5\}) = 14.5$. Then, $\tilde{v}(S) + \tilde{v}(T) > \tilde{v}(S \cup T) + \tilde{v}(S \cap T)$. So, $\tilde{v}$ is not convex.

Table 1. A weighted bankruptcy situation and the corresponding WCEA allocation.

$i \in N$	c_i (CO2)	a_i (GDP)	λ_i	x_i (WCEA)
1	9	3	3	$\frac{3}{2}$
2	12	6	2	3
3	5	5	1	$\frac{5}{2}$
4	4	10	$\frac{2}{5}$	4
5	3	9	$\frac{1}{3}$	3

The previous example shows that we cannot immediately guarantee the balancedness of TT- games using the convexity argument (and, similarly, we cannot use this argument to guarantee that the Shapley value lies the core of TT-games). Nevertheless, in the following we introduce a rule providing an allocation always in the core of a TT-game.

Definition 2. Let $(N, \tilde{v})$ be the TT-game corresponding to (E, c, a).
Let $i^* \in \arg\min_{i \in N : x_i - c_i = 0} \lambda_i$. Define the allocation $z \in \mathbb{R}^N$ such that for each $i \in N$

$$z_i = \begin{cases} \frac{x_i}{\lambda(N \setminus \{i^*\})} - \frac{x_i}{\lambda_i} & \text{if } c_i - x_i > 0, \\ 0 & \text{if } c_i - x_i = 0 \text{ and } i \neq i^*, \\ \tilde{v}(N) - \sum_{i \in N \setminus \{i^*\}} z_i & \text{if } i \neq i^*. \end{cases} \quad (5)$$

Proposition 2. *TT-games are balanced.*

Proof. Consider a TT-game $(N, \tilde{v})$ corresponding to the allocation generated by the weighted CEA rule $WCEA(E, c, a) = (x_1, x_2, \ldots, x_n)$. Allocation z is clearly efficient, as $\sum_{i \in N} z_i = \tilde{v}(N)$. Let $P = \{i \in N : c_i - x_i > 0\}$, and let $I = N \setminus P$. If $P \neq \emptyset$, from the fact that $\lambda(N \setminus \{i^*\}) \leq \lambda_i$, we have that $z_i \geq 0$ for each $i \in P$. Moreover, by the fact $\lambda(N \setminus \{i^*\}) \geq \lambda(N) = \lambda_{i^*}$, we also have that

$$\tilde{v}(N) = \sum_{i \in P} \frac{x_i}{\lambda_{i^*}} - \frac{x_i}{\lambda_i} \geq \sum_{i \in P} \left(\frac{x_i}{\lambda(N \setminus \{i^*\})} - \frac{x_i}{\lambda_i} \right),$$

and then it immediately follows that $z_{i^*} = \tilde{v}(N) - \left(\frac{x_i}{\lambda(N \setminus \{i^*\})} - \frac{x_i}{\lambda_i} \right) \geq 0$. Then, for each $S \subseteq N$ such that $S \cap P = \emptyset$ or $S \cap I = \emptyset$, we have that $\sum_{i \in S} z_i \geq 0 = \tilde{v}(S)$.

Now, let $S \subseteq N$ be such that $S \cap P \neq \emptyset$ and $S \cap I \neq \emptyset$. If $i^* \notin S$, we have that

$$\tilde{v}(S) = \sum_{i \in S \cap P} \frac{x_i}{\lambda(S)} - \frac{x_i}{\lambda_i} \leq \sum_{i \in S \cap P} \frac{x_i}{\lambda(N \setminus \{i^*\})} - \frac{x_i}{\lambda_i} = \sum_{i \in S} z_i, \qquad (6)$$

where the first equality follows directly from relation (4), and the inequality follows from the fact that $\lambda(S) \geq \lambda(N \setminus \{i^*\})$. Otherwise, if $i^* \in S$, we have that

$$\tilde{v}(S) = \sum_{i \in S \cap P} \frac{x_i}{\lambda(S)} - \frac{x_i}{\lambda_i} \leq \sum_{i \in S \cap P} \frac{x_i}{\lambda(N \setminus \{i^*\})} - \frac{x_i}{\lambda_i} + z_{i^*} = \sum_{i \in S} z_i, \qquad (7)$$

where the inequality follows from the previous arguments and the fact that we have shown earlier that $z_{i^*} \geq 0$. So, we have proved that $z \in C(\tilde{v})$.

5   Conclusions

In this paper, we have studied how to allocate CO_2 emissions permits among the EU-28 using bankruptcy rules and we have shown that the WCEA rule keeps into account not only countries' claims but also countries' productions. In order to study incentives to the transfer of efficient technologies from countries with a low emission intensity to those with a high one, we have proposed a preliminary approach based on cooperative games and we have shown that, thanks to the properties of the allocation provided by the weighted CEA rule, the core of these games is non-empty. For space reasons, we have omitted a more exhaustive analysis of the WCEA solution in comparison with other allocation methods for CO_2 permits proposed in the literature. We finally observe that the GDP index is often considered as a good measure to compare the environmental performance of countries with similar social and economic conditions. However, for scenarios dealing with heterogeneous states, the use of a more comprehensive estimation of the environmental performance of countries policies is recommended.

References

1. Berk, M.M., den Elzen, M.G.: Options for differentiation of future commitments in climate policy: how to realise timely participation to meet stringent climate goals? Clim. Policy **1**(4), 465–480 (2001)
2. Carraro, C., Eyckmans, J., Finus, M.: Optimal transfers and participation decisions in international environmental agreements. Rev. Int. Organ. **1**(4), 379–396 (2006)
3. Casas-Mẃndez, B., Fragnelli, V., García-Jurado, I.: Weighted bankruptcy rules and the museum pass problem. Eur. J. Oper. Res. **125**(1), 161–168 (2011)
4. Eyckmans, J., Tulkens, H.: Simulating coalitionally stable burden sharing agreements for the climate change problem. In: Chander, P., Drèze, J., Lovell, C.K., Mintz, J. (eds.) Public Goods, Environmental Externalities and Fiscal Competition, pp. 218–249. Springer, Boston (2006). https://doi.org/10.1007/978-0-387-25534-7_13
5. Filar, J.A., Gaertner, P.S.: A regional allocation of world CO2 emission reductions. Math. Comput. Simul. **43**(3–6), 269–275 (1997)
6. Giménez-Gómez, J.M., Teixidó-Figueras, J., Vilella, C.: The global carbon budget: a conflicting claims problem. Clim. Change **136**(3–4), 693–703 (2016)
7. Gomes, E.G., Lins, M.E.: Modelling undesirable outputs with zero sum gains data envelopment analysis models. J. Oper. Res. Soc. **59**(5), 616–623 (2008)
8. Gutiérrez, E., Llorca, N., Sánchez-Soriano, J., Mosquera, M.: Sustainable allocation of greenhouse gas emission permits for firms with Leontief technologies. Eur. J. Oper. Res. **269**(1), 5–15 (2018)
9. Grubb, M.: The Greenhouse Effect: Negotiating Targets, p. 60. Royal Institute of International Affairs, London (1989)
10. Mianabadi, H., Mostert, E., Pande, S., van de Giesen, N.: Weighted bankruptcy rules and transboundary water resources allocation. Water Resour. Manage. **29**(7), 2303–2321 (2015)
11. Mianabadi, H., Mostert, E., Zarghami, M., van de Giesen, N.: A new bankruptcy method for conflict resolution in water resources allocation. J. Environ. Manage. **144**, 152–159 (2014)
12. Moulin, H.: Priority rules and other asymmetric rationing methods. Econometrica **68**(3), 643–684 (2000)
13. O'Neill, B.: A problem of rights arbitration from the Talmud. Math. Soc. Sci. **2**(4), 345–371 (1982)
14. Rose, A., Stevens, B., Edmonds, J., Wise, M.: International equity and differentiation in global warming policy. Environ. Resour. Econ. **12**(1), 25–51 (1998)
15. Schmidt, R.C., Heitzig, J.: Carbon leakage: grandfathering as an incentive device to avert firm relocation. J. Environ. Econ. Manage. **67**(2), 209–223 (2014)
16. Thomson, W.: Axiomatic and game-theoretic analysis of bankruptcy and taxation problems: a survey. Math. Soc. Sci. **45**(3), 249–297 (2003)
17. Welsch, H.: A CO2 agreement proposal with flexible quotas. Energy Policy **21**(7), 748–756 (1993)
18. Wood, P.J.: Climate change and game theory. Ann. New York Acad. Sci. **1219**(1), 153–170 (2011)
19. Zarezadeh, M., Mirchi, A., Read, L., Madani, K.: Ten bankruptcy methods for resolving natural resource allocation conflicts. Water Diplomacy Action: Conting. Approach. Manag. Complex Water Probl. **1**, 37 (2017)

20. Zhu-Gang, J., Wen-Jia, C., Can, W.: Simulation of climate negotiation strategies between China and the US based on game theory. Adv. Clim. Change Res. **5**(1), 34–40 (2014)
21. Zhou, P., Wang, M.: Carbon dioxide emissions allocation: a review. Ecol. Econ. **125**, 47–59 (2016)

The Economics of Bundling Content with Unlicensed Wireless Service

Yining Zhu[(✉)], Haoran Yu, and Randall Berry

Department of Electrical and Computer Engineering, Northwestern University,
Evanston, USA
`yiningzhu2015@u.northwestern.edu`, `yhrhawk@gmail.com`,
`rberry@eecs.northwestern.edu`

Abstract. Adding new unlicensed wireless spectrum is a promising approach to accommodate increasing traffic demand. However, unlicensed spectrum may have a high risk of becoming congested, and service providers (SPs) may have difficulty to differentiate their wireless services when offering them on the same unlicensed spectrum. When SPs offer identical services, the resulting competition can lead to zero profits. In this work, we consider the case where an SP bundles its wireless service with a content service. We show that this can differentiate the SPs' services and lead to positive SP profits. In particular, we study the characteristics of the content services that an SP should bundle with its wireless service, and analyze the impact of bundling on consumer surplus.

Keywords: Unlicensed spectrum market · Game theory · Bundling

1 Introduction

Motivated in part by the success of WiFi, there is an increasing interest in expanding the amount of spectrum available for unlicensed access. Having new unlicensed spectrum can increase competition in the wireless service market and promote the development of new technologies. In addition to the TV white spaces [1] and the Generalized Authorized Access (GAA) tier in the 3.5 GHz band [2], the FCC in the U.S. has recently proposed opening up the 6 GHz band for unlicensed use [3].

Adding unlicensed spectrum promotes competition in that a service provider (SP) can enter the market without paying a license cost. However, the open access of unlicensed spectrum may also lead it to be overcrowded, which results in a "tragedy of commons." Moreover, as shown in [10,11], competition among SPs in an unlicensed band may result in a "price war," in which no SP makes any profit from the unlicensed wireless service.

To avoid the price war, we propose a market strategy in which an SP can bundle its wireless service with a content service. Based on [6–9], commodity

This research was supported in part by NSF grants TWC-1314620, AST- 1343381, AST-1547328 and CNS-1701921.

K. Avrachenkov et al. (Eds.): GameNets 2019, LNICST 277, pp. 93–108, 2019.
https://doi.org/10.1007/978-3-030-16989-3_7

bundling is a prevalent marketing strategy that brings cost and information advantages. Furthermore, it enables providers to sort customers into groups with different reservation prices and extract consumer surplus. According to [9], bundling typically reduces the diversity of reservation prices of consumers, and thereby enables sellers to extract more consumer surplus. In this paper, we want to investigate whether a similar phenomenon will happen in a wireless market, which differs from a commodity market in that the SPs utilize a congestible resource to offer service. Furthermore, the reason for zero SP profits in the unlicensed spectrum market is that the SPs offer identical service [10,11]. Bundling the wireless service with other commodities can differentiate the SPs' services, and potentially lead to positive SP profits. The bundling can be realized through mergers and acquisitions. For example, AT&T has completed the acquisition of Time Warner Inc., and now offers an unlimited data plan bundled with HBO (one of time Warner's leading video services) to its wireless customers [4,5]. Bundling can also be realized by the cooperation of multiple subsidiaries from the same conglomerate. For example, Google owns both Project-Fi providing wireless service and YouTube Red, a content service.

There have been many recent works studying the competition among wireless service providers on an unlicensed band, e.g., [10–15]. In [10,11,14], models of price competition with unlicensed service were studied, which we will adopt in this work. As we have already discussed, [10,11] showed that price competition can result in zero profits. Reference [12] proposed using contracts to reduce the risk of losing revenue for an incumbent SP on an unlicensed band. Based on [12], reference [13] considered the investment and technology upgrade decisions of new entrant SPs. Reference [15] proposed an alternative market structure based on short-term permits, which was shown to achieve positive profits. As in [12–15], our approach of bundling provides another way to sustain positive profits and is more in line with current practices in the wireless market. We also study the impact of bundling on consumer surplus.

The main questions we want to answer in this paper are as follows:

– Is bundling a promising strategy to use in a wireless market with unlicensed spectrum? Can the SPs achieve an equilibrium that leads to positive profits?
– What are the characteristics (e.g., popularity and value) of the content services that an SP should consider while making the bundling decision?
– How does bundling affect the consumer surplus?

In this paper, we consider two SPs and build a three-stage Stackelberg game to model the bundling decision and competition between the SPs. The main results are as follows:

– The market equilibrium exists and the SPs' profits will increase if an SP bundles its wireless service with a suitable content service.
– The SP should bundle with a content service whose popularity is below an upper bound and value is above a lower bound. Among the services that satisfy these bounds, the SP should choose the one with a high popularity to increase its profit.

– Customer surplus will decrease when an SP chooses to bundle unless the band resource is extremely limited.

2   Model

We use B to denote the unlicensed bandwidth and parameter $x_p < B$ to denote the background traffic on the unlicensed band. We assume that there are two service providers, i.e., SP1 and SP2, who offer wireless service using the unlicensed band. SP1 has the option to bundle its wireless service with the content service, which can differentiate its service from that of SP2. In this work, we do not consider the case where SP2 also has a bundling option. If SP2 bundles its wireless service with the same content service, both SPs still offer the same type of service and achieve zero profits under price competition. If SP2 bundles its wireless service with a different content service, the problem's analysis depends on the customers' valuations on the two content service and we leave such analysis as future work. We formulate the SPs' interactions as the following three-stage Stackelberg game:

– Stage I: SP1 decides whether to bundle its wireless service with the content service. Moreover, SP1 announces p_c, which is the content service's retail price.
– Stage II: SP1 and SP2 decide their prices p_1 and p_2 for wireless customers. If SP1 chooses bundling, p_1 is the price for the bundled service (which contains both the wireless and content service); otherwise, it is the price for the wireless service. Note that p_2 is always SP2's wireless service price.
– Stage III: The customers decide which services to subscribe to, based on p_1, p_2, p_c, and SP1's bundling decision.

2.1   Content Service

In this paper, we assume that the cost of providing the content service does not change with the number of customers using it. Examples are content services provided by companies like HBO, TIDAL and Netflix. We consider two types of customers, who have different valuations for the content service. A fraction α of the customers have high valuations, i.e., θ_h, and a fraction $1 - \alpha$ of the customers have low valuations, i.e., θ_l. We assume that $\alpha \theta_h \geq \theta_l$. Intuitively, this means the content service can generate more profit when the content service provider chooses a high price to serve only the customers with high valuations instead of choosing a low price to serve all customers.

2.2   Wireless Service

As in [10, 11, 13], the SPs compete for a common pool of customers to maximize their profits. The customers are modeled as non-atomic users with a total mass of 1. Each customer may choose an SP considering its *delivered price*, which is the sum of the actual price paid and a congestion cost. We use x_1 to denote the total mass of customers that subscribe to SP1's service, and x_2 to denote the total mass of customers that subscribe to SP2's service. When the total traffic on the unlicensed band is x_T, the congestion cost of using the unlicensed band is $\frac{x_T}{B}$. This implies that the congestion is linearly increasing in the traffic [12,13]. The total traffic x_T includes the background traffic x_p as well as the traffic of the SPs' customers. Then, the delivered price of SP1, denoted as y_1, is $\frac{x_T}{B} + p_1$. Similarly, we use $y_2 = \frac{x_T}{B} + p_2$ to denote the delivered price of SP2.

Each customer is identified as $x \in [0,1]$. We use $v(x)$ to denote the customer's valuation for the wireless service, and assume that it is given by (similar assumptions have been made in [12,13]):

$$v(x) = 1 - x.$$

We use θ_x to denote customer x's valuation for the content service. It is a random variable, whose value is given by

$$\theta_x = \begin{cases} \theta_h, & \text{w.p. } \alpha, \\ \theta_l, & \text{w.p. } 1 - \alpha. \end{cases}$$

Recall that α is the fraction of customers with high valuations for the content service.

The traffic generated by a customer on the unlicensed band depends on whether it uses the content service. If a customer does not use the content service, the generated traffic is normalized to 1; otherwise, its generated traffic is $\beta \geq 1$. Note that $\beta > 1$ models a case where subscribing to the content service increases the amount of traffic a customer consumes.

2.3 Customers' Choice

We use a vector $\mathbf{d} \triangleq (d_1, d_2, s)$ to denote a customer's service subscription decision. If the customer subscribes to SP1's service, $d_1 = 1$; otherwise, $d_1 = 0$. If the customer subscribes to SP2's wireless service, $d_2 = 1$; otherwise, $d_2 = 0$. If the customer subscribes to the content service, $s = 1$; otherwise, $s = 0$. A customer decides $\mathbf{d}$ to maximize its welfare. We denote SP1's bundling decision by $b \in \{0, 1\}$, where $b = 1$ means bundling. Thus, if $b = 0$, customer x's welfare given $\mathbf{d}$ would be

$$CW(x, b, \mathbf{d}) = v(x) \cdot \max\{d_1, d_2\} + \theta_x \cdot s - y_1 \cdot d_1 - y_2 \cdot d_2 - p_c \cdot s,$$

where $v(x) \cdot \max\{d_1, d_2\}$ captures that the customer's utility includes $v(x)$ if and only if it subscribes to the wireless service.

If $b = 1$, customer x's welfare is given by

$$CW(x, b, \mathbf{d}) = v(x) \cdot \max\{d_1, d_2\} + \theta_x \cdot \max\{d_1, s\} - y_1 \cdot d_1 - y_2 \cdot d_2 - p_c \cdot s,$$

where $\theta_x \cdot \max\{d_1, s\}$ captures that the customer can use the content service by two different approaches when SP1 chooses bundling. First, the customer can subscribe to SP1's (bundled) service, i.e., $d_1 = 1$. Second, the customer can subscribe to the content service alone, i.e., $s = 1$. It is easy to see that a customer will not choose $d_1 = 1$ and $s = 1$ at the same time when SP1 chooses bundling. Given SP1's bundling decision b, customer x will make decisions to maximize his welfare. Thus, its optimal choice, denoted as $d^*(x, b)$, is given by:

$$d^*(x, b) = \arg\max_{d \in \{0,1\}^3} CW(x, b, d). \tag{1}$$

3   Wardrop Equilibrium in Stage III

In Sect. 2.3, we discussed an individual customer's choice given SP1's bundling decision, the service prices, and the congestion. In this section, we study the customers' optimal decisions in Stage III and the resulting equilibrium market shares of the SPs. Specifically, we consider the Wardrop Equilibrium [17]. The basic intuition of Wardrop Equilibrium in our model is that, at Wardrop Equilibrium,

- for the customers who only have wireless services, the delivered price they pay should be the same;
- for the customers who have both content service and wireless service, the delivered price they pay should be the same.

Based on the Wardrop Equilibrium conditions, we can analyze the mass of customers choosing different services in equilibrium. Next, we discuss the cases where $b = 0$ and $b = 1$, separately.

3.1   Benchmark Case ($b = 0$)

We denote the case where SP1 does not bundle ($b = 0$) as the *benchmark case*, and the case where SP1 bundles its wireless service with the content service ($b = 1$) as the *bundling case*. In the benchmark case, a customer subscribes to the content service if and only if the content service price p_c is no greater than its valuation. Based on our assumption $\alpha\theta_h > \theta_l$, we can see that SP1 should choose $p_c = \theta_h$ to maximize its profit generated from the content service. Each customer decides which SP's service to subscribe to based on p_1, p_2, and the congestion cost. If $p_1 = p_2$, the Wardrop equilibrium is given by

$$\begin{cases} \frac{\beta\alpha(x_1+x_2)+(1-\alpha)(x_1+x_2)+x_p}{B} + p_1 = 1 - x_1 - x_2, \\ \frac{\beta\alpha(x_1+x_2)+(1-\alpha)(x_1+x_2)+x_p}{B} + p_2 = 1 - x_1 - x_2, \end{cases} \tag{2}$$

where the total traffic x_T equals $\beta\alpha(x_1 + x_2) + (1 - \alpha)(x_1 + x_2) + x_p$ and $\frac{\beta\alpha(x_1+x_2)+(1-\alpha)(x_1+x_2)+x_p}{B}$ is the congestion cost. When SP1 chooses $p_c = \theta_h$, only the customers with high valuations subscribe to the content service. Hence, among the $x_1 + x_2$ customers using the wireless service, a fraction α of them will use the content service on the unlicensed band. Each of them generates β traffic on the unlicensed band. Note that among the $1 - x_1 - x_2$ customers who do not use the wireless service, a fraction α of them also subscribe to the content service. These customers can use the content service via other approaches (e.g., wire-line networks) and do not generate traffic on the unlicensed band. If $p_1 > p_2$, the equilibrium is given by

$$
\begin{cases}
x_1 = 0, \\
\frac{\beta \alpha x_2 + (1-\alpha) x_2 + x_p}{B} + p_2 = 1 - x_2.
\end{cases}
\tag{3}
$$

Similarly, if $p_1 < p_2$, the equilibrium is given by

$$
\begin{cases}
x_2 = 0, \\
\frac{\beta \alpha x_1 + (1-\alpha) x_1 + x_p}{B} + p_1 = 1 - x_1.
\end{cases}
\tag{4}
$$

When $p_1 = p_2$, the delivered prices of SP1's and SP2's services are equal. In this case, each customer whose reservation price is greater than the delivered price subscribes to SP1's or SP2's services with an equal probability. An SP can always choose a price that is slightly smaller than the other SP's price to capture the entire market. As a result, this again leads to a price war between the two SPs: they will set prices to zero and have zero profits from the wireless market at the equilibrium (though SP 1 does still have its profit from the content service).

3.2   Bundling Case ($b = 1$)

By considering $CW(x, b, \mathbf{d})$ when $b = 1$, we summarize a customer's optimal choice as follows. Customer x will subscribe to SP1's bundled service ($\mathbf{d}^*(x, 1) = \{1, 0, 0\}$) if

$$
\begin{cases}
y_1 \le y_2 + \min\{p_c, \theta_x\}, \\
y_1 \le 1 - x + \min\{p_c, \theta_x\}.
\end{cases}
\tag{5}
$$

Recall that y_1 and y_2 are the delivered prices of SP1 and SP2, which are the sum of the congestion cost of the unlicensed spectrum and the service price of each SP. This means that a customer with value $\theta_x > p_c$ will subscribe to SP1's service if the delivered price for SP1's bundled service is smaller than (i) the sum of SP2's delivered price and p_c and (ii) the sum of the customer's value of wireless service and p_c. For a customer with value $\theta_x < p_c$, it will subscribe to SP1's service if the delivered price for SP1's bundled service is smaller than (i) the sum of SP2's delivered price and θ_x and (ii) the sum of the customer's value of wireless service and θ_x.

Similarly, the customer will subscribe to SP2's service and the content service separately ($\mathbf{d}^*(x, 1) = \{0, 1, 1\}$) if

$$
\begin{cases}
\theta_x \ge p_c, \\
y_1 > y_2 + p_c, \\
y_2 \le 1 - x.
\end{cases}
\tag{6}
$$

The customer will subscribe to the content service only $(\mathbf{d}^*(x,1) = \{0,0,1\})$ if

$$
\begin{cases}
\theta_x \geq p_c, \\
y_1 > 1 - x + p_c, \\
y_2 > 1 - x.
\end{cases}
\tag{7}
$$

The customer will subscribe to SP2's service only $(\mathbf{d}^*(x,1) = \{0,1,0\})$ if

$$
\begin{cases}
\theta_x < p_c, \\
y_1 > y_2 + \theta_x, \\
y_2 \leq 1 - x.
\end{cases}
\tag{8}
$$

The customer will not subscribe to any service $(\mathbf{d}^*(x,1) = \{0,0,0\})$ if

$$
\begin{cases}
\theta_x < p_c, \\
y_1 > 1 - x + \theta_x, \\
y_2 > 1 - x.
\end{cases}
\tag{9}
$$

As we can observe from the conditions above, the customer's choice of services is determined by the delivered prices and the retail price of the content service. There are three possible ranges of p_c (i.e., the content service's price): (a) $\theta_l < p_c \leq \theta_h$, (b) $p_c \leq \theta_l$, (c) $p_c > \theta_h$. We first analyze case (a), and we will later show that SP1 will not choose p_c to be in cases (b) or (c) in equilibrium. Note that in case (a), only the customers with high valuations will pay p_c to subscribe to the content service.

Given that $\theta_l < p_c \leq \theta_h$, we next discuss four cases for the difference between SP1's and SP2's delivered prices: (i) $y_1 - y_2 \leq \theta_l$, (ii) $\theta_l < y_1 - y_2 < p_c$, (iii) $y_1 - y_2 = p_c$ and (iv) $y_1 - y_2 > p_c$. Note that $y_1 - y_2$ is equivalent to $p_1 - p_2$, since the two SPs use the same unlicensed band to serve customers. We assume that $\theta_l = 0$ to simplify the analysis in the rest of the paper.

Case (i) $(y_1 - y_2 \leq \theta_l)$: From (6) and (8), we can see that $d_2 = 1$ only if $y_1 - y_2 > \min\{p_c, \theta_x\}$. Hence, if $y_1 - y_2 \leq \theta_l$, no customer will subscribe to SP2's service. From (5), we can see that the customers with $v(x) \geq y_1$ will subscribe to SP1's service regardless of their θ_x. Customers with $v(x) \in [\max\{0, y_1 - p_c\}, y_1)$ will subscribe to SP1's service only if $\theta_x > p_c$. For the customers with $x \in (1 - y_1, \min\{1, 1 - y_1 + p_c\}]$, a fraction α of them subscribe to SP1's service.

As $\theta_l < p_c \leq \theta_h$ is assumed in this case, α fraction of the customers in $(1 - y_1, \min\{1, 1 - y_1 + p_c\}]$ join SP1. Thus, the Wardrop Equilibrium in this case is given by

$$
\begin{cases}
p_1 < p_2, \\
x_2 = 0, \\
y_1 = \frac{\beta x_1 + x_p}{B} + p_1, \\
x_1 = (1 - y_1) + \alpha \min\{p_c, y_1\}.
\end{cases}
\tag{10}
$$

Case (ii) ($\theta_l < y_1 - y_2 < p_c$): The condition $y_1 - y_2 > \theta_l$ implies $y_1 > y_2 + \min\{p_c, \theta_x\}$ when $\theta_x = \theta_l$. From (5), we can see that the customers with low valuations for the content service will not subscribe to SP1. However, customers will subscribe to SP1's service if $x \leq \min\{1, 1 - y_1 + p_c\}$ and $\theta_x = \theta_h$. We can also see that the customers will subscribe to SP2's service if $x \leq 1 - y_2$ and $\theta_x < y_1 - y_2$. Since $\theta_l \leq y_1 - y_2 < p_c \leq \theta_h$, the condition for a customer to subscribe to SP2's service reduces to $x \leq 1 - y_2$ and $\theta_x = \theta_l$. This indicates that for the customers who subscribe to wireless services, customers with higher value of the content service will subscribe to SP1's service and customers with lower value of the content service will subscribe to SP2's service. As a result, we can compute SP1's and SP2's market shares as $x_1 = \alpha(\min\{1, 1 - y_1 + p_c\})$, and $x_2 = (1 - \alpha)(1 - y_2)$. The Wardrop Equilibrium in this case is given by

$$
\begin{cases}
0 < p_1 - p_2 < p_c, \\
\frac{\beta x_1 + x_2 + x_p}{B} + p_2 = 1 - x_1 - x_2, \\
x_1 = \alpha(1 - \max\{0, \frac{\beta x_1 + x_2 + x_p}{B} + p_1 - p_c\}), \\
x_2 = (1 - \alpha)(1 - \min\{1, \frac{\beta x_1 + x_2 + x_p}{B} + p_2\}).
\end{cases}
\tag{11}
$$

Case (iii) ($y_1 - y_2 = p_c$): SP1 and SP2 have the same delivered price for the wireless and content service combination. For the customers with $\theta_x = \theta_h$ and $x \leq 1 - y_2$, they can either (i) subscribe to SP1's bundled service or (ii) subscribe to SP2's service and SP1's content service separately, which lead to the same welfare for the customers. From (8), we can see that the customers with $v(x) \leq y_2$ (i.e., $x \leq 1 - y_2$) will subscribe to SP2's service. We use x_{21} to denote the total number of customers who subscribe to both SP2's service and SP1's content service. Moreover, we use x_{20} to denote the total number of customers who only subscribe to SP2's service (without SP1's content service). Since x_2 is SP2's market share, we have $x_2 = x_{20} + x_{21}$. Based on the analysis above, x_1, x_{21}, and x_{20} satisfy $x_{21} + x_1 = \alpha(1 - y_2)$ and $x_{20} = (1 - \alpha)(1 - y_2)$. The Wardrop Equilibrium in this case is given by

$$
\begin{cases}
p_1 = p_2 + p_c, \\
\frac{\beta(x_1 + x_{21}) + x_{20} + x_p}{B} + p_2 = 1 - x_1 - x_{21} - x_{20}, \\
x_1 = x_{21}, \\
x_{20} = (1 - \alpha)(x_1 + x_{21} + x_{20}).
\end{cases}
\tag{12}
$$

The condition $x_1 = x_{21}$ implies that when (i) subscribing to SP1's bundled service and (ii) subscribing to both SP2's service and SP1's content service generate the same welfare, a customer will randomly choose one of these two options with an equal probability.

Case (iv) ($y_1 - y_2 > p_c$): From (5), we can see that no customer will subscribe to SP1's bundled service. Thus, the Wardrop Equilibrium is given by

$$
\begin{cases}
p_1 > p_2 + p_c, \\
x_1 = 0, \\
\frac{\beta \alpha x_2 + (1 - \alpha) x_2 + x_p}{B} + p_2 = 1 - x_2.
\end{cases}
\tag{13}
$$

Combining the analysis of Wardrop Equilibrium in cases (i)–(iv), we can compute SP1 and SP2's market shares as functions of their prices:

$$x_1(p_1, p_2, p_c) = \tag{14}$$

$$
\begin{cases}
\frac{B(1-(1-\alpha)p_1)-(1-\alpha)x_p}{(1-\alpha)\beta+B}, & \text{if } p_1 \leq \min\{p_2, \bar{p_1}^1\}, \\
\frac{B(1-p_1+\alpha p_c)-x_p}{\beta+B}, & \text{if } \min\{p_2, \bar{p_1}^1\} \leq p_1 \leq p_2, \\
\frac{\alpha(B(1-p_1+p_c)-x_p)}{\alpha\beta+B}, & \text{if } p_2 < p_1 \leq \min\{\max\{p_2, \bar{p_1}^2\}, \bar{p_1}^3\}, \\
\alpha, & \text{if } \min\{\max\{p_2, \bar{p_1}^2\}, \bar{p_1}^3\} < p_1 \leq \max\{p_2, \bar{p_1}^2\}, \\
\frac{a((-1+\alpha)(p_1-p_2-p_c)+B(1-p_1+p_c)-x_p)}{1+\alpha(\beta-1)+B}, & \text{if } \max\{p_2, \bar{p_1}^2\} < p_1 < p_2 + p_c, \\
\frac{\alpha(B(1-p_2)-x_p)}{2(1+\alpha(\beta-1)+B)}, & \text{if } p_1 = p_2 + p_c, \\
0, & \text{if } p_1 > p_2 + p_c;
\end{cases}
$$

$$x_2(p_1, p_2, p_c) = \tag{15}$$

$$
\begin{cases}
\frac{B-Bp_2-x_p}{1+\alpha(\beta-1)+B}, & \text{if } p_2 < p_1 - p_c, \\
\frac{(2-\alpha)(B(1-p_2)-x_p)}{2(1+\alpha(\beta-1)+B)}, & \text{if } p_2 = p_1 - p_c, \\
\frac{(1-\alpha)(B(1-p_2)+\alpha\beta(p_1-p_2-p_c)-x_p)}{1+\alpha(\beta-1)+B}, & \text{if } p_1 - p_c < p_2 \leq \min\{\max\{\bar{p_2}^1, p_1 - p_c\}, \bar{p_2}^2\}, \\
-\frac{(1-\alpha)(\alpha\beta-B(1-p_2)+x_p)}{1-\alpha+B}, & \text{if } \max\{\bar{p_2}^1, p_1 - p_c\} < p_2 < \min\{p_1, \bar{p_2}^3\}, \\
0, & \text{otherwise.}
\end{cases}
$$

Here, $\bar{p_1}^1 \triangleq \frac{-\beta+(1-\alpha)\beta p_c + Bp_c - x_p}{B}$, $\bar{p_1}^2 \triangleq -\frac{1-p_2-p_c-Bp_c+\alpha(\beta-1+p_2+p_c)+x_p}{1-\alpha+B}$, $\bar{p_1}^3 \triangleq \frac{B(-1+p_2)+\alpha\beta(p_2+p_c)+x_p}{\alpha\beta}$, $\bar{p_2}^1 \triangleq \frac{1+p_1+Bp_1-p_c-Bp_c+a(\beta-1-p_1+p_c)+x_p}{1-\alpha}$, $\bar{p_2}^2 \triangleq \frac{B+\alpha\beta(p_1-p_c)-x_p}{B+\alpha\beta}$, and $\bar{p_2}^3 \triangleq 1 - \frac{\alpha\beta+x_p}{B}$.

4   Equilibrium Analysis for Stages II and I

In Sect. 3, we obtained the market shares of the SPs given the prices and bundling choice. Based on this, we can further analyze SP1's and SP2's pricing and bundling decisions. Assuming that SP1 bundles its service, we first compute their profits given the pricing decisions as follows:

$$\pi_1(p_1, p_2, p_c) =$$

$$
\begin{cases}
x_1(p_1, p_2, p_c) \cdot p_1 + \alpha \cdot \max\{0, y_1(p_1, p_2, p_c) - p_c\} \cdot p_c, & p_1 \leq p_2, \\
x_1(p_1, p_2, p_c) \cdot p_1 + (\alpha - x_1(p_1, p_2, p_c)) \cdot p_c, & p_2 < p_1 < p_2 + p_c, \\
x_1(p_1, p_2, p_c) \cdot p_1 + (\alpha - 2x_1(p_1, p_2, p_c)) \cdot p_c, & p_1 = p_2 + p_c, \\
\alpha \cdot p_c, & p_1 > p_2 + p_c;
\end{cases}
$$
$$\tag{16}$$

$$\pi_2(p_1, p_2, p_c) = x_2(p_1, p_2, p_c) \cdot p_2. \tag{17}$$

When SP1 does not bundle its wireless service with the content service, there is a unique price equilibrium in Stage II, which leads to zero wireless service profits for the SPs. When SP1 chooses bundling, there might not exist a price equilibrium in Stage II. In this case, it would be difficult for SP1 to estimate its profit under the bundling choice. Hence, we assume that SP1 will not choose bundling if there does not exist a price equilibrium in Stage II. Next, we show that when SP1 chooses bundling, there exists a price equilibrium in Stage II if and only if the content service's characteristics satisfy certain conditions. Moreover, we will show that when these conditions are satisfied, SP1 should choose bundling in Stage I, which improves SP1's total profit over the benchmark case. We introduce the following theorem.

Theorem 1. *When $b = 1$, there exists a unique price equilibrium in Stage II if and only if the following conditions hold:*

$$\begin{cases} \theta_h > \max\{\frac{(1-\alpha+\alpha\beta)(B-x_p)}{3\alpha\beta(1-\alpha)+4B(1+B-\alpha+\alpha\beta)}, \bar{p}_c\}, \\ \alpha < \frac{1}{1+\beta}, \\ \frac{(1+\alpha(2\beta-1)+2B)}{3\alpha\beta(1-\alpha)+4B(1+B-\alpha+\alpha\beta)} < \frac{1-\sqrt{\alpha}}{\alpha\beta(1+\sqrt{\alpha})+2B}. \end{cases} \tag{18}$$

Here, $\bar{p}_c$ is the larger solution of the two solutions of the quadratic equation[1] $\pi_1(p_2^{PE}(p_c), p_2^{PE}(p_c), p_c) = \pi_1(p_1^{PE}(p_c), p_2^{PE}(p_c), p_c)$.
When these conditions hold, the equilibrium prices are given by:

$$p_c = \theta_h, \tag{19}$$

$$p_1 = \theta_h + \frac{(1+\alpha(2\beta-1)+2B)(B-x_p)}{3\alpha\beta(1-\alpha)+4B(1+B-\alpha+\alpha\beta)}, \tag{20}$$

$$p_2 = \frac{(\alpha\beta+2B+2(1-\alpha))(B-x_p)}{3\alpha\beta(1-\alpha)+4B(1+B-\alpha+\alpha\beta)}, \tag{21}$$

Moreover, when the conditions in (18) hold, SP1 should choose bundling in Stage I, i.e., $b^* = 1$.

We use Fig. 1 to illustrate how the price equilibrium is achieved. The left figure in Fig. 1 shows how SP1's profit changes with p_1 when $p_2 = p_2^*$, and the right one shows how SP2's profit changes with p_2 when $p_1 = p_1^*$. We can observe that both p_1^* and p_2^* are the global optimal solutions. Thus, the price equilibrium is achieved, and it is in Wardrop Equilibrium case (ii).

5   Types of Content Service

For both the benchmark case and bundling case, we have derived the price equilibrium between SP1 and SP2. We can compare SP1's profits under the benchmark case and bundling case, and then analyze the conditions under which bundling improves SP1's profit. This will imply what types of content service SP1 should consider for bundling.

---

[1] We use the superscript PE to indicate that the corresponding results are derived at price equilibrium.

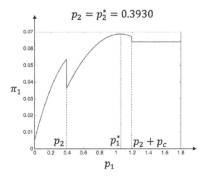
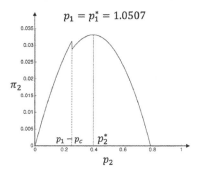

Fig. 1. An example of how price equilibrium is achieved when $B = 0.2$, $\alpha = 0.08$, $x_p = 0$, $\beta = 1$ and $p_c = \theta_h = 0.8$.

In our model, we consider three dimensions of the content service, its popularity (α), value (θ_h) and congestion factor (β). In the conditions needed for SP1 to bundle in (18), these three parameters are closely related to each other. To get a clearer intuition, we analyze these three dimensions separately and focus on one of them in each subsection.

5.1 Popularity

From (18), if θ_h is large enough, α needs to be smaller than a certain upper bound, which is affected by β and B. In Fig. 2, we give the region of α and B that satisfies the feasible condition when $\beta = 1$. This figure shows that, when the band resource (B) increases, SP1 should consider less popular services to bundle. SP1 hopes that with bundling, SP2 will not lower its price to start a price war. Note that when SP1 chooses bundling, the services of SP1 and SP2 are differentiated. In this case, SP2 may also attract customers (i.e., those with low valuations for the content service) and get a positive profit. If α is large, most of the customers have high valuations for the video service and SP2 can hardly attract any customers. This gives SP2 motivation to lower its price, which starts a price war and leads to zero profits to both SPs. The following lemma gives a more general upper bound of α for all possible B.

Lemma 1. *SP1 should bundle its wireless service with a content service when the content service has a popularity less than $\bar{\alpha}$, where $\bar{\alpha} = \min\{\frac{5+2\beta-3\sqrt{1+4\beta}}{2(-2+\beta)^2},$ $\frac{1}{1+\beta}\}$.*

Note that this result gives a loose upper bound and does not depend on the unlicensed bandwidth B and background traffic x_p. For any given B and x_p, interestingly, SP1 should not bundle its wireless service with a very popular content service. Moreover, given any β, α needs to be smaller than 0.146, which is because $\bar{\alpha} < 0.146$. To get a better intuition of the actual value of α in practice, we give a numerical example here. There are around 400 million wireless

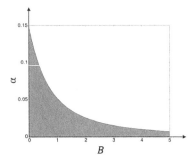

Fig. 2. The region of α and B that satisfies the feasible condition when $\beta = 1$.

subscribers in the U.S. Suppose that there are 600 million potential users (in our model, they correspond to all customers who need to make decisions). According to [16], the video subscription company with the largest market share (i.e., Netflix) has an α of around 0.088, and the company with the second largest market share (i.e., Amazon) has an α of around 0.043. Some content service providers have lower popularity, e.g., HBO Now and YouTube Red have an α of around 0.0083 and 0.0025, respectively.

5.2 Value

From (18), we observe that θ_h needs to be greater than a lower bound. We give an example to show how this bound changes with the popularity α, the band resource B and the amount of primary traffic in the unlicensed band x_p in Fig. 3. The figure shows that, this bound is increasing with B and decreasing with α and x_p. The reasons are as follows. Bundling helps prevent the price war in the sense that if SP1 lowers its price of the bundled service to attract all customers in the wireless market, it will lose a large profit generated from offering the content service. If θ_h is high enough, SP1 will not make its price lower than SP2's price, because the fraction of customers with high valuations for the content service is large and reducing price will greatly reduce SP1's profit from the content service. If B increases, the profit from the wireless market increases, which requires a larger θ_h to hedge SP1's desire to lower its price. On the other hand, the increase of x_p decreases the profit from the wireless market, which in turn lowers the requirement for θ_h. If α increases, the profit from the content service increases, which in turn eases the requirement for θ_h.

5.3 Congestion Factor

From (18), we observe that when β increases, the upper bound of α and the lower bound of θ_h decrease. This means that among bandwidth-consuming services, e.g., video services, the content services that SP1 should bundle its wireless service with should have a larger range of values and a smaller range of popularity.

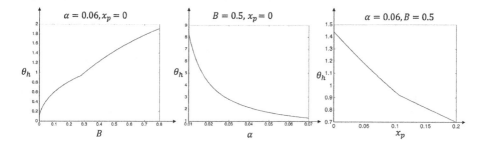

Fig. 3. Impact of α, x_p and B on θ's upper bound when $\beta = 1$.

The reason is that, increasing β decreases the profit from the wireless service, which eases the requirement for θ_h. Increasing β also implies more competition in the wireless market. Thus, a smaller α is required to avoid SP2 decreasing its price.

6 Profits

In this section, we analyze SP1's and SP2's profits in the price equilibrium, compare them in the bundling case and benchmark case, and investigate the impact of the content service's characteristics on the SPs' profits. We first introduce the following theorem.

Theorem 2. *If the conditions in (18) hold, both SP1 and SP2 achieve higher profits in the bundling case, compared to the benchmark case.*

This can be verified by substituting (19) into (16), and comparing it with the SP profits in the benchmark case. We give an example of profit comparison in Fig. 4. In the left figure, the solid lines are the profits of SP1 and SP2 in the bundling case, and the blue dashed line is profit of SP1 in the benchmark case. Note that in the benchmark case, SP1 only generates profit from the content service, and cannot generate profit from the wireless service. It can be observed that bundling improves both SP1's and SP2's profits. In particular, both SPs can generate positive profits from the wireless service. From the left figure, we can also see that SP1's profit in the bundling case increases with α. This implies that SP1 should choose a service with α as large as possible in the feasible region determined by (18).

In the right figure, we compare the increase in profits because of bundling. We define the profit increase of SP1 as $\pi_1(p_1^{PE}(\theta_h), p_2^{PE}(\theta_h), \theta_h) - \alpha\theta_h$ and the profit increase of SP2 as $\pi_2(p_1^{PE}(\theta_h), p_2^{PE}(\theta_h), \theta_h)$ (SP2 has zero profit in the benchmark case). It can be proved that the profit increase of SP1 is always smaller than the profit increase of SP2 in the feasible region, and we can also observe this from the right figure. The insight is that, both SP1 and SP2 might prefer to wait for the other SP to bundle if they both have the bundling option. However, if both of them wait and do not bundle, they will get zero profit from wireless service.

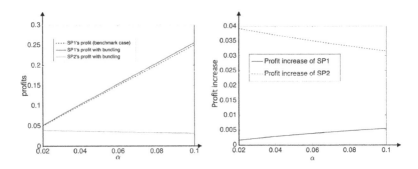

Fig. 4. Profits when $B = 0.2$, $\beta = 1$, and $\theta_h = 2.5$.

Theorem 3. *If conditions in (18) hold, $\pi_1(p_1^{PE}(\theta_h), p_2^{PE}(\theta_h), \theta_h)$ is increasing in θ_h, but the profit increase of SP1 is independent of θ_h.*

This theorem indicates that SP1's total profit in the bundling case is increasing in θ_h, which implies that SP1 prefers a content service with high value. However, a high value of θ_h also improves SP1's profit without bundling. The net effect of this is that the increase in profit does not depend on θ_h.

Lemma 2. *If conditions in (18) hold, $\pi_1(p_1^{PE}(\theta_h), p_2^{PE}(\theta_h), \theta_h)$ is decreasing in β.*

This indicates that SP1 should prefer the service with a smaller congestion factor in the feasible region defined in (18). This is intuitive as bundling improves the mass of customers using the content service and generates more congestion on the unlicensed band. This decreases the profits of the SPs. One example is that, when a video subscription service and a music subscription service have similar popularity and values, an SP might prefer the latter service to bundle with.

7   Consumer Surplus

In this section, we compare the consumer surplus, which is the integral of all customers' welfare, in the benchmark case and bundling case. As shown in Fig. 5, bundling decreases the consumer surplus in most cases, expect for the cases when B is extremely small. In the left side of Fig. 5, we consider an extremely small B, i.e., $B = 0.02$. In this case, when α is large, the consumer surplus can be increased by bundling. The intuition is that when the band resource is extremely limited, bundling helps reduce the competition and decrease the congestion, so that consumer surplus increases. However, this effect will be negligible if α is too small and SP2 almost takes the whole market. When B is bigger ($B = 0.07$ in the right figure), bundling always decreases the consumer surplus. When B is large, the consumer surplus in the benchmark case will be much larger than that in the bundling case.

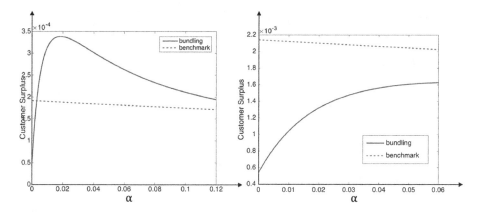

Fig. 5. Two examples of customer surplus. Left: $B = 0.02$ and $\beta = 1.5$; Right: $B = 0.07$ and $\beta = 1.5$.

8   Conclusion

In this paper, we considered the use of bundling as a means of forming a niche unlicensed wireless service market. We studied a case where two SPs compete on an unlicensed band. We showed that an SP can bundle its wireless service with a content service to differentiate the SPs' services. We proved that if the content service has a low popularity, a high value, and a small congestion factor, there exists a unique price equilibrium. Moreover, in this case, both SPs can achieve positive profits and there is no price war between them. We also showed that bundling decreases the consumer surplus, except for the extreme case where the band resource is very limited.

References

1. Federal Communications Commission: Unlicensed operation in the TV broadcast bands/additional spectrum for unlicensed devices below 900 MHz and in the 3 GHz band, FCC Report and Order, September 2010
2. Federal Communications Commission: Amendment of the commission's rules with regard to commercial operations in the 3550–3650 MHz band, Report and order and second further notice of proposed rulemaking (2015)
3. Federal Communications Commission: Promoting unlicensed use of the 6 GHz band, Notice of Proposed Rulemaking ET Docket No. 18–295; GN Docket No. 17–183, October 2018
4. https://about.att.com/story/att_completes_acquisition_of_time_warner_inc.html
5. https://www.att.com/plans/unlimited-data-plans.html
6. Adams, W.J., Yellen, J.L.: Commodity bundling and the burden of monopoly. Q. J. Econ. **90**(3), 475–498 (1976)
7. Bakos, Y., Brynjolfsson, E.: Bundling information goods: pricing, profits, and efficiency. Manag. Sci. **45**(12), 1613–1630 (1999)

8. Chuang, J.C.-I., Sirbu, M.A.: Optimal bundling strategy for digital information goods: network delivery of articles and subscriptions. Inf. Econ. Policy **11**(2), 147–176 (1999)
9. Schmalensee, R.: Gaussian demand and commodity bundling. J. Bus. **57**(1), S211–S230 (1984)
10. Nguyen, T., Zhou, H., Berry, R.A., Honig, M.L., Vohra, R.: The impact of additional unlicensed spectrum on wireless services competition. In: Proceedings of IEEE DySPAN, Aachen, Germany (2011)
11. Nguyen, T., Zhou, H., Berry, R.A., Honig, M.L., Vohra, R.: The cost of free spectrum. Oper. Res. **64**(6), 1217–1229 (2017)
12. Zhu, Y., Berry, R.A.: Contract as entry barriers in unlicensed spectrum. In: IEEE INFOCOM Workshop on Smart Data Pricing, May 2017
13. Zhu, Y., Berry, R.A.: Contracts as investment barriers in unlicensed spectrum. In: Proceedings of IEEE INFOCOM, Hawaii, U.S. (2018)
14. Zhang, F., Zhang, W.: Competition between wireless service providers: pricing, equilibrium and efficiency. In: Proceedings of IEEE WiOpt, Tsukuba Science City, Japan (2013)
15. Wang, X., Berry, R.A.: The impact of short-term permits on competition in unlicensed spectrum. In: Proceedings of IEEE DySPAN (2017)
16. https://www.statista.com/statistics/185390/leading-cable-programming-networks-in-the-us-by-number-of-subscribers/
17. Smith, M.J.: The existence, uniqueness and stability of traffic equilibria. Transp. Res. Part B Methodol. **13**(4), 295–304 (1979)

Normalized Equilibrium in Tullock Rent Seeking Game

Eitan Altman[1,2,3(✉)], Mandar Datar[1,2], Gerard Burnside[1,3],
and Corinne Touati[1]

[1] Joint Lab of INRIA and Nokia Bell-Labs, Paris, France
{eitan.altman,mandar.datar,corinne.touati}@inria.fr
[2] Laboratoire Informatique d'Avignon, University of Avignon, Avignon, France
[3] Laboratory of Information, Networking and Communication Sciences (LINCS),
Paris, France
gerard.burnside@nokia-bell-labs.com

Abstract. Games with Common Coupled Constraints represent many real-life situations. In these games, if one player fails to satisfy its constraints common to other players, then the other players are also penalized. Therefore these games can be viewed as being cooperative in goals related to meeting the common constraints, and non-cooperative in terms of the utilities. We study in this paper the Tullock rent-seeking game with additional common coupled constraints. We have succeeded in showing that the utilities satisfy the property of diagonal strict concavity (DSC), which can be viewed as an extension of concavity to a game setting. It not only guarantees the uniqueness of the Nash equilibrium but also of the normalized equilibrium.

Keywords: Normalized equilibrium · Common Coupled Constraints ·
Diagonal strict concavity

1 Introduction

Games with constraints have long been used for modeling and studying non-cooperative behavior in various areas. This includes road traffic [7,12] and telecommunications [9]. Various types of constraints may appear in everyday game situations; the simplest ones consisting of orthogonal constraints, where the strategies of the players are restricted independently of each other [15]. The second type of constraints are called Common Coupled Constraints (CCC) [3,14,15] in which all players have a common convex non-orthogonal multi-strategy space. This model can be viewed as constraints that are common to all users. A unilateral deviation of a player from some feasible multi-strategy (one that satisfies the constraints) to another strategy that is feasible for that player, does not result, therefore, in the violation of constraints of other users. CCC has often been used in telecommunications networking problems as well as in power transfer over a smart grid, where capacity constraints of links are naturally common. Games

© ICST Institute for Computer Sciences, Social Informatics and Telecommunications Engineering 2019
Published by Springer Nature Switzerland AG 2019. All Rights Reserved
K. Avrachenkov et al. (Eds.): GameNets 2019, LNICST 277, pp. 109–116, 2019.
https://doi.org/10.1007/978-3-030-16989-3_8

with this type of constraints are a special case of General Constrained Games (GCG) [6], see also [3–5,8,10,16].

In this paper, we study the well known Tullock rent-seeking game with Common Coupled Constraints. This game describes contest over resources. Each player bids an amount that she is ready to pay. She then pays an amount proportional to her bid and receives, in turn, a payoff that is proportional to her bid divided by the sum of bids of all players.

The presence of a capacity constraint results in infinitely many equilibria and we are faced with a question of equilibrium selection. Using Kuhn Tucker conditions to the best response, we can solve a relaxed game instead of the original constrained game, which has, however, the same equilibria as the original game. The Lagrange multipliers can be interpreted as a shadow cost that a manager sets in order to guarantee that the equilibrium achieved satisfies the constraints. This approach may, however, be completely unscalable since KKT Theorem does not guarantee that the price per resource unit is the same for all players. In fact, since the Lagrange multipliers are obtained for the best response function, they could depend not only on the player but also on the policy of all other players, rending the approach even less scalable. We are interested in finding such shadow cost which is fixed per resource unit. Such an equilibrium along with a fixed shadow price is called a normalized equilibrium.

The Tullock rent-seeking game has been used recently to model and study several game phenomena in networking. It was used to model contests over timelines in social networks for maximizing visibility [17]. Each player i controls the rate $\lambda_i a_i$ of a Poisson process of posts that player i sends into a common timeline of length K. This rate is given by a basic popularity rate λ_i times the acceleration effort (e.g. through advertisement) given by a_i. Using basic queueing theory, the authors show that the stationary expected number of posts in the timeline originating from player i is given by

$$K \frac{\lambda_i a_i}{\sum_{j=1}^{N} \lambda_j a_j}$$

This visibility measure is the payoff in Tullock's model, while the cost for acceleration at a rate a_i is proportional to a_i as in Tullock's model.

Another application of the Tullock rent-seeking game is in the study of contests between miners in blockchain [2].

A few words on rent-seeking. According to Wikipedia, "In public choice theory and in economics, rent-seeking involves seeking to increase one's share of existing wealth without creating new wealth. Rent-seeking results in reduced economic efficiency through the poor allocation of resources, reduced actual wealth-creation, lost government revenue, increased income inequality, and (potentially) national decline."

Wikipedia further describes the origin of the idea: "The idea of rent-seeking was developed by Gordon Tullock in 1967, while the expression rent-seeking itself was coined in 1974 by Krueger [11]. The word "rent" does not refer specifically to payment on a lease but rather to Adam Smith's division of incomes into profit, wage, and rent. The origin of the term refers to gaining control of land or other natural resources."

Our first contribution is to show that the utilities satisfy a property that extends concavity to games, and is called Diagonally Strict Concavity. This is shown to imply the existence and uniqueness of a normalized equilibrium. We shall then show that this property further extends to the case of contests over several resources.

2   A Single Resource

Consider an N players game. Player m bids a quantity x^m. We have minimum constraints $x^m \geq \epsilon$ for all m.

The payoff from this contest to player m is

$$P^m = \frac{x^m}{\sum_{j=1}^{M} x^j}.$$

This comes at a cost of $x^m \gamma$ to player m where γ is a constant. The utility for player m is thus

$$U^m(x) = \frac{x^m}{\sum_{j=1}^{M} x^j} - x^m \gamma.$$

Theorem 1. *(i) The utility of player m is concave in its action and is continuous in the actions of other players.*
(ii) For any strictly positive value of γ, the above game has a unique Nash equilibrium in pure policies.

Proof. Direct calculation leads to (i). The existence then directly follows from [15]. Uniqueness is established in [1], see also [18]. Other related uniqueness results in the asymmetric case can be found in [17,19].

3   Normalized Equilibrium

The games we have seen so far involved orthogonal constraints. By that, we mean that the actions that a player can use do not depend on the actions of other players. We next introduce capacity constraint. We require that the following holds for some constant V:

$$\sum_{j=1}^{M} x^j \leq V \tag{1}$$

Capacity constraints may represent physical bounds on resources, such as bounded power, or resources that are bounded by regulation. For example, legislation may impose bounds on the power used or on the emission of CO_2 by cars. With the additional capacity constraint, the Nash equilibrium is no more unique and there may, in fact, be an infinite number of equilibria. We call this the game with capacity constraint.

Let y be an equilibrium in the above game and let $y_{[-m]}$ denote the action vectors of all players other than m. By KKT Theorem, since for each m, U^m is concave in x^m, there is a Lagrange multiplier $\lambda^m(y_{[-m]})$ such that y^m maximizes the Lagrangian

$$L^m(x^m) = U^m(x, y_{[-m]}) - \lambda^m(y_{[-m]}) \left(V_k - \sum_{j=1}^{M} x^j \right)$$

and

$$\lambda^m(y_{[-m]}) \left(V - \sum_{j=1}^{M} x^j \right) = 0$$

(complementarity property). We call the game with the Lagrangian L^m replacing the utilities U^m the relaxed game.

The Lagrange multipliers can be interpreted as shadow prices: if a price is set on player m such that when other players are at equilibrium, the player pays $x^m \lambda^m(y_{[-m]})$ for its use of the capacity, then y is an equilibrium in the game with capacity constraints. Yet this pricing is not scalable since for the same use of the resources it may vary from user to user and it further depends on the chosen equilibrium. For billing purposes, one would prefer λ^m not to depend on y nor on m, but to be a constant.

Does there exist a constant Lagrange multiplier λ independent of strategies of the payers and of the identity m of the player, along with an associated equilibrium y for the corresponding relaxed game? If the answer is positive then y is called a *normalized equilibrium* [15].

Our goal is to establish the existence and uniqueness of the normalized equilibrium.

4   Diagonal Strict Concavity

For a vector of real non-negative numbers r, define

$$\sigma(x, r) = \sum_{m=1}^{N} r_m U^m(x)$$

$$g(x, r) = \begin{bmatrix} r_1 \frac{\partial}{\partial x_1} U^1(x_1, x_{-1}) \\ r_2 \frac{\partial}{\partial x_2} U^2(x_2, x_{-2}) \\ \vdots \\ r_N \frac{\partial}{\partial x_N} U^N(x_N, x_{-N}) \end{bmatrix} \tag{2}$$

σ is called diagonally strict concave (DSC) for a given r if for every distinct x^0 and x^1,

$$(x^1 - x^0)' g(x^0, r) + (x^0 - x^1)' g(x^1, r) > 0$$

Let $G(x,r)$ be the Jacobian of $g(x,r)$ with respect to x and let $G_{i,j}$ be i^{th} row and j^{th} column of $G(x,r)$. Then a sufficient condition for σ to be diagonally strict concave for some r is that for all x, $[G(x,r)+G'(x,r)]$ is negative definite.

Our interest in diagonally strict concave utility functions is due to the following properties of games possessing such utilities.

Theorem 2 *(Theorem 4 from [15]). Let σ be diagonally strict concave for some r. Then there exists a unique normalized equilibrium.*

5   Proof of DSC

In this section we establish that the Tullock game with capacity constraint has a DSC structure and thus has a unique normalized equilibrium.

In our case we have

$$
g(x,r) = \begin{bmatrix} \frac{r_1 x_{-1}}{X} \\ \frac{r_2 x_{-2}}{X} \\ \vdots \\ \frac{r_N x_{-N}}{X} \end{bmatrix}
\tag{3}
$$

where $X = \sum_{i=1}^{N} x_i$ and $x_{-m} = \sum_{i=1,i\neq m}^{N} x_i$

$$
G_{i,j} = \frac{\partial}{\partial x_j}\left(\frac{\partial}{\partial x_i} \frac{r_i x_i}{X} \right)
\tag{4}
$$

$$
r_i \frac{\partial}{\partial x_j}\left(\frac{x_{-i}}{X^2} \right) = \begin{cases} r_i \frac{-2x_i}{X^3} & \text{if } i = j \\ r_i \frac{x_i - x_{-i}}{X^3} & \text{if } i \neq j \end{cases}
\tag{5}
$$

For $[G+G']$ consider

$$
G_{i,j} + G_{j,i} = \begin{cases} \frac{-4r_i x_{-i}}{X^3} & \text{if } i = j \\ \frac{r_i(x_i - x_{-i}) + r_j(x_j - x_{-j})}{X^3} & \text{if } i \neq j \end{cases}
\tag{6}
$$

$[G+G']$ is negative definite if $A'[G+G']A < 0, \forall A, A \neq 0$ where A is the column vector

$$
A = \begin{bmatrix} a_1 \\ \vdots \\ a_N \end{bmatrix}
\tag{7}
$$

$$
A'[G+G']A
$$

$$
= \sum_{i=1}^{N}\left[\sum_{j=1,j\neq i}^{N} a_i a_j \frac{r_i(x_i - x_{-i}) + r_j(x_j - x_{-j})}{X^3} \right] - a_i^2 \frac{4r_i x_{-i}}{X^3}
$$

We choose $r_i = 1$ for all i. Then (7) equals $-Z/X^3$ where Z is given by

$$\sum_{i=1}^{N} a_i^2 4x_{-i} + \left[\sum_{j=1,j\neq i}^{N} a_i a_j \left((x_{-i} - x_i) + (x_{-j} - x_j) \right) \right] \tag{8}$$

$$= \sum_{i=1}^{N} a_i^2 4 \left(X - x_i \right) + \left[\sum_{j>i}^{N} 4 a_i a_j \left(X - x_i - x_j \right) \right] \tag{9}$$

$$= 4 \sum_{i=1}^{N} a_i^2 \left(X - x_i \right) + \left[\sum_{j>i}^{N} a_i a_j \left(X - x_i - x_j \right) \right] \tag{10}$$

$$= 4 \sum_{i=1}^{N} \left[a_i^2 \sum_{j=1,j\neq i}^{N} x_i + \sum_{j>i}^{N} a_i a_j \sum_{k=1,k\neq j,k\neq i}^{N} x_k \right] \tag{11}$$

$$= \sum_{i=1}^{N} 4x_i \left[\sum_{j=1,j\neq i}^{N} a_j^2 + a_j \sum_{k>j,k\neq i}^{N} a_k \right] \tag{12}$$

Now (12) is positive for any positive value of x and hence [G'+G] matrix is negative definite.

6   Several Resources

We consider next the following extension to the case of K resources. Each player m of the M players has a budget $B(m)$ that he can invest by bidding x_k^m of resource k. The following (orthogonal) constraint should hold:

$$\sum_{k=1}^{K} x_k^m \leq B(m).$$

The payoff for player m is the sum of payoffs in all K contests, i.e.

$$P^m(x) = \sum_{k=1}^{K} P_k^m(x_k)$$

where x_k is the vector $x_k^1, \ldots, x_k^M$ and where

$$P_k^m(x_k) = \frac{x_k^m}{\sum_{j=1}^{M} x_k^j}.$$

and the cost of a contest k to player m is $\gamma(k)x_k^m$. Player m's utility is thus

$$U^m(x) = \sum_{k=1}^{K} \left(\frac{x_k^m}{\sum_{j=1}^{M} x_k^j} - \gamma_k x_k^m \right)$$

For the study of such games, see [17].

We next define capacity constraint on each of the K resources. Let V be the column vector with the kth entry being a constant V_k. We then require for each k that

$$\sum_{m=1}^{N} x_k^m \leq V_k$$

Note that when applying KKT conditions to the best response at equilibrium, we shall have K Lagrange multipliers. We wish to find a vector of K Lagrange multipliers that do not depend on the player nor on the policy of other players, such that the Nash equilibrium for the relaxed game will be an equilibrium for the original constrained game and in particular the constraints would be met and would satisfy the complementarity conditions. This is the vector version of a normalized equilibrium.

According to Theorem 4 of Rosen [15], we have to show that the set of utilities is diagonally strict concave in order to have existence and uniqueness of the normalized equilibrium. The latter follows from the fact that DSC holds for each resource separately and then apply the proof of Corollary 2 in [13].

7   Conclusions and Future Work

We have shown that the utilities in the Tullock game are strict diagonal concave. This allows to conclude using Rosen's result that in absence of common correlated constraints, the Nash equilibrium exists and is unique, while in presence of such constraints, the normalized equilibrium exists in pure strategies and is unique. Note that while the statements on the Nash equilibrium have already been available in [1] which proposed an extension to the DSC property, that reference does provide tools to handle the normalized equilibrium.

Another advantage from the derivation of the DSC structure is that one can use dynamic distributed algorithms to converge to the normalized equilibrium and convergence is guaranteed under DSC, see [15].

References

1. Altman, E., Hanawal, M.K., Sundaresan, R.: Generalising diagonal strict concavity property for uniqueness of Nash equilibrium. Indian J. Pure Appl. Math. **47**(2), 213–228 (2016). https://doi.org/10.1007/s13226-016-0185-4. hal-01340963

2. Altman, E., Reiffers-Masson, A., Menasché, D.S., Datar, M., Dhamal, S., Touati, C.: Mining competition in a multi-cryptocurrency ecosystem at the network edge: a congestion game approach. In: 1st Symposium on Cryptocurrency Analysis, SOCCA 2018, Toulouse, France, December 2018, pp. 1–4 (2018)

3. Altman, E., Solan, E.: Constrained games: the impact of the attitude to adversary's constraints. IEEE Trans. Autom. Control **54**(10), 2435–2440 (2009). https://doi.org/10.1109/TAC.2009.2029302

4. Altman, E., Shwartz, A.: Constrained Markov games: nash equilibria. In: Filar, J.A., Gaitsgory, V., Mizukami, K. (eds.) Advances in Dynamic Games and Applications. Annals of the International Society of Dynamic Games, vol. 5. Birkhauser, Boston (2000)
5. Altman, E., Avrachenkov, K., Bonneau, N., Debbah, M., El-Azouzi, R., Menasche, D.S.: Constrained cost-coupled stochastic games with independent state processes. Oper. Res. Lett. **36**, 160–164 (2008)
6. Debreu, G.: A social equilibrium existence theorem. Proc. National Acad. Sci. U.S.A. **38**, 886–893 (1952)
7. El-Azouzi, R., Altman, E.: Constrained traffic equilibrium in routing. IEEE Trans. Autom. Control **48**, 1656–1660 (2003)
8. Gomez-Ramırez, E., Najim, K., Poznyak, A.S.: Saddle-point calculation for constrained finite Markov chains. J. Econ. Dyn. Control **27**, 1833–1853 (2003)
9. Hsiao, M.T., Lazar, A.A.: Optimal decentralized flow control of Markovian queueing networks with multiple controllers. Perform. Eval. **13**, 181–204 (1991)
10. Kallenberg, L.C.M.: Linear Programming and Finite Markovian Control Problems, p. 148. Mathematical Centre Tracts, Amsterdam (1983)
11. Krueger, A.: The political economy of the rent-seeking society. Am. Econ. Rev. **64**(3), 291–303 (1974). JSTOR 1808883
12. Larsson, T., Patriksson, M.: Side constrained traffic equilibrium models - traffic management through link tolls. In: Marcotte, P., Nguyen, S. (eds.) Equilibrium and Advanced Transportation Modelling, pp. 125–151. Kluwer Academic Publishers (1998)
13. Orda, A., Rom, R., Shimkin, N.: Competitive routing in multi-user communication networks. IEEE/ACM Trans. Netw. **1**(5), 510–521 (1993)
14. Pavel, L.: An extension of duality to a game-theoretic framework. Automatica **43**(2), 226–237 (2007)
15. Rosen, J.: Existence and uniqueness of equilibrium points for concave N-person games. Econometrica **33**, 520–534 (1965)
16. Shimkin, N.: Stochastic games with average cost constraints. In: Annals of the International Society of Dynamic Games, vol. 1. Birkhauser, Boston (1994)
17. Reiffers-Masson, A., Hayel, Y., Altman, E.: Game theory approach for modeling competition over visibility on social networks. In: 2014 6th International Conference on Communication Systems and Networks, COMSNETS 2014, pp. 1–6 (2014). https://doi.org/10.1109/COMSNETS.2014.6734939
18. Jensen, M.K.: Existence, uniqueness and comparative statics in contests. University of Leicester, Working Paper No. 15/16, July 2015
19. Szidarovszky, F., Okuguchi, K.: On the existence and uniqueness of pure nash equilibrium in rent-seeking games. Games Econ. Behav. **18**(1), 135–140 (1997)

Game Theory for Social Networks

Community Structures in Information Networks

Martin Carrington and Peter Marbach$^{(\boxtimes)}$

Department of Computer Science, University of Toronto, Toronto, Canada
{carrington,marbach}@cs.utoronto.edu

Abstract. We study community structures that emerge in an information network using a game-theoretic approach. In particular, we consider a particular family of community structures, and provide conditions under which there exists a Nash equilibrium within this family.

Keywords: Social networks · Information networks · Community structure

1 Introduction

In this paper we consider a particular type of social network, which we refer to as an *information network*, where agents (individuals) share/exchange information. Sharing/exchanging of information is an important aspect of social networks, both for social networks that we form in our everyday lives, as well as for online social networks such as Twitter.

The work in [1] presents a model to study communities in information networks where agents produce (generate) content, and consume (obtain) content. Furthermore, the model allows agents to form communities in order to share/exchange content more efficiently, where agents obtain a certain utility for joining a given community. Using a game-theoretic framework, [1] characterizes the community structures that emerge in information networks as Nash equilibria. More precisely, [1] considers a particular family of community structures, and shows that (under suitable assumptions) there always exists a community structure that is a Nash equilibrium. One open question from [1] is whether the family of community structures considered includes all Nash equilibria, or whether there exist Nash equilibria that are not covered by the analysis in [1]. In this paper we address this question, and show that there do indeed exist Nash equilibria that are not covered by the analysis in [1]. One interesting, and important, characteristic is that the Nash equilibria that we derive in this paper have the property that some agents (individuals) are "excluded" from the community structure, i.e. do not participate in any of the information communities. If such Nash equilibria are to emerge in real-life (social) information networks, it would mean that some individuals are "marginalized". This is definitely an undesirable

© ICST Institute for Computer Sciences, Social Informatics and Telecommunications Engineering 2019
Published by Springer Nature Switzerland AG 2019. All Rights Reserved
K. Avrachenkov et al. (Eds.): GameNets 2019, LNICST 277, pp. 119–127, 2019.
https://doi.org/10.1007/978-3-030-16989-3_9

outcome that could come at great cost for the individuals that are "marginal-ized". As such, understanding when the Nash equilibria obtained in this paper do emerge in (social) information networks is an important question.

The rest of the paper is organized as follows. In Sect. 2 we summarize the model presented in [1] that we use for our analysis. In Sect. 3 we define the family of community structures that we consider in this paper, and in Sect. 4 we present our results.

Due to space constraints we refer to [1] for a review of related literature, and we only point to the work on content forwarding and filtering in social networks by Zadeh, Goel and Munagala [2], and the work by Hegde, Massoulie, and Vien-not [3], as they are most closely related to the analysis presented in this paper. In [2], Zadeh, Goel and Munagala consider the problem of information diffusion in social networks under a broadcast model where content forwarded (posted) by a user is seen by all its neighbors (followers, friends) in the social graph. For this model, the paper [2] studies whether there exists a network structure and filtering strategy that lead to both high recall and high precision. High recall means that all users receive all the content that they are interested in, and high precision means that all users only receive content they are interested in. The main result in [2] shows that this is indeed the case under suitable graph models such as for example Kronecker graphs. In [3], Hegde, Massoulie, and Viennot study the problem where users are interested in obtaining content on specific topics, and study whether there exists a graph structure and filtering strategy that allows users to obtain all the content they are interested in. Using a game-theoretic framework (flow games), the analysis in [3] shows that under suitable assumptions there exists a Nash equilibrium, and selfish dynamics converge to a Nash equilibrium. The main difference between the model and analysis in [2,3] and the approach in this paper is that model and analysis in [2,3] does not explicitly consider and model community structures, and the utility obtained by users under the models in [2,3] depends only on the content that agents receive, but not on the content agents produce.

2   Background

In this section we review the model and results of [1]. Due to space constraints we keep the presentation of the model brief, and refer to [1] for a more detailed discussion of the model, and the results that were obtained in [1]. For our analy-sis we assume that each content item that is being produced in the information community is of a particular type. One might think of a content type as a topic, or an interest, that agents might have. Furthermore we assume that there exists a structure that relates different content types to each other. In particular, we assume there exists a measure of "closeness" between content types that charac-terizes how strongly related two content types are. For example, as "basketball" and "baseball" are both sports one would assume that the two topics are more closely related than "basketball" and "mathematics". To model this situation we assume that the type of a content item is given by a point x in a metric space, and the closeness between two content types $x, x' \in \mathcal{M}$ is then given by the distance measure $d(x, x')$, $x, x' \in \mathcal{M}$, for the metric space $\mathcal{M}$.

Having defined the set of content that can be produced in an information network, we next describe agents' interests in content as well as the agents' ability to produce content. To do this, we assume that there is a set $\mathcal{A}_d$ of agents that consume content, and a set $\mathcal{A}_s$ of agents that produce content, where the subscripts stand for "demand" and "supply". Furthermore, we associate with each agent that consumes content a center of interest $y \in \mathcal{M}$, i.e. the center of interest y of the agent is the content type (topic) that an agent is most interested in. The interest in content of type x of an agent with center of interest y is given by

$$p(x|y) = f(d(x, y)), \qquad x, y \in \mathcal{M}, \tag{1}$$

where $d(x, y)$ is the distance between the center of interest y and the content type x, and $f : [0, \infty) \mapsto [0, 1]$ is a non-increasing function. The interpretation of the function $p(x|y)$ is as follows: when an agent with center of interest y consumes (reads) a content item of type x, then it finds it interesting with probability $p(x|y)$ as given by Eq. (1). As the function f is non-increasing, this model captures the intuition that the agent is more interested in content that is close to its center of interest y.

Similarly, given an agent that produces content, the center of interest y of the agent is the content type (topic) that the agent is most adept at producing. The ability of the agent to produce content of type $x \in \mathcal{M}$ is then given by

$$q(x|y) = g(d(x, y)), \tag{2}$$

where $g : [0, \infty) \mapsto [0, 1]$ is a non-increasing function.

In the following we identify an agent by its center of interest $y \in \mathcal{M}$, i.e. agent y is the agent with center of interest y. As a result we have that $\mathcal{A}_d \subseteq \mathcal{M}$ and $\mathcal{A}_s \subseteq \mathcal{M}$.

2.1   Information Community

We model an information community as follows. An information community $C = (C_d, C_s)$ consists of a set of agents that consume content $C_d \subseteq \mathcal{A}_d$ and a set of agents that produce content $C_s \subseteq \mathcal{A}_s$. Let $\beta_C(x|y)$ be the rate at which agent $y \in C_s$ generates content items of type x in community C. Let $\alpha_C(y)$ be the fraction of content produced in community C that agent $y \in C_d$ consumes. To define the utility for content consumption and production, we assume that when an agent consumes a single content item, it receives a reward equal to 1 if the content item is of interest and relevant, and pays a cost of $c > 0$ for consuming the item. The cost c captures the cost in time (energy) to read/consume a content item. Using this reward and cost structure, the utility rate ("reward minus cost") for content consumption of agent $y \in C_d$ is given by (see [1] for a detailed derivation)

$$U_C^{(d)}(y) = \alpha_C(y) \int_{x \in \mathcal{M}} [Q_C(x)p(x|y) - \beta_C(x)c]dx,$$

where

$$Q_C(x) = \int_{y \in C_s} \beta_C(x|y) q(x|y) dy, \quad \text{and} \quad \beta_C(x) = \int_{y \in C_s} \beta_C(x|y) dy.$$

Similarly, the utility rate for content production of agent $y \in C_s$ is given by

$$U_C^{(s)}(y) = \int_{x \in \mathcal{M}} \beta_C(x|y)[q(x|y)P_C(x) - \alpha_C c] dx,$$

where

$$P_C(x) = \int_{y \in C_d} \alpha_C(y) p(x|y) dy, \quad \text{and} \quad \alpha_C = \int_{z \in C_d} \alpha_C(z) dz.$$

As discussed in [1], the utility rate for content production captures how "valuable" the content produced by agent y is for the set of content consuming agents C_d in the community C.

2.2   Community Structure and Nash Equilibrium

Using the above definition of a community, a community structure that describes how agents organize themselves into communities is then given by a triplet $(\mathcal{C}, \{\alpha_C(y)\}_{y \in \mathcal{A}_d}, \{\beta_C(\cdot|y)\}_{y \in \mathcal{A}_s})$, where the set of communities $\mathcal{C}$ in this structure consists of communities C as defined in the previous section, and

$$\alpha_C(y) = \{\alpha_C(y)\}_{C \in \mathcal{C}}, y \in \mathcal{A}_d, \quad \text{and} \quad \beta_C(\cdot|y) = \{\beta_C(\cdot|y)\}_{C \in \mathcal{C}}, y \in \mathcal{A}_s,$$

are the consumption fractions and production rates, respectively, that agents allocate to the different communities $C \in \mathcal{C}$. We assume that the total consumption fractions and production rates of each agent are bounded by $E_p > 0$, and $E_q > 0$, respectively, i.e. we have that

$$\|\alpha_C(y)\| = \sum_{C \in \mathcal{C}} \alpha_C(y) \le E_p \le 1, \qquad y \in \mathcal{A}_d,$$

and

$$\|\beta_C(y)\| = \sum_{C \in \mathcal{C}} \|\beta_C(\cdot|y)\| \le E_q, \qquad y \in \mathcal{A}_s,$$

where

$$\|\beta_C(\cdot|y)\| = \int_{x \in \mathcal{M}} \beta_C(x|y) dx.$$

We assume that agents form communities in order to maximize their utility rates, i.e. agents join communities, and choose allocations $\alpha_C(y)$, and $\beta_C(\cdot|y)$ to maximize their total consumption, and production utility rates, respectively.

A Nash equilibrium is then given by a community structure $(\mathcal{C}^*, \{\alpha_C^*(y)\}_{y \in \mathcal{A}_d}, \{\beta_C^*(\cdot|y)\}_{y \in \mathcal{A}_s})$ such that for all agents $y \in \mathcal{A}_d$ we have that

$$\alpha_C^*(y) = \operatorname*{arg\,max}_{\alpha_C(y): \|\alpha_C(y)\| \le E_p} \sum_{C \in \mathcal{C}} U_C^{(d)}(y),$$

and for all agents $y \in \mathcal{A}_s$, we have that

$$\beta_{\mathcal{C}}^*(\cdot|y) = \underset{\beta_{\mathcal{C}}(\cdot|y):\|\beta_{\mathcal{C}}(y)\| \leq E_q}{\arg\max} \sum_{C \in \mathcal{C}} U_C^{(s)}(y).$$

We call a Nash equilibrium a covering Nash equilibrium if for all agents $y \in \mathcal{A}_d$, we have that there exists at least one community $C \in \mathcal{C}$ such that $\alpha_C(y) > 0$, and for all agents $y \in \mathcal{A}_d$, we have that there exists at least one community $C \in \mathcal{C}$ such that $\|\beta_C(\cdot|y)\| > 0$.

2.3   Results

The above model has been analyzed in [1] for the case of a specific metric space, and a specific family of information communities. More precisely, the analysis in [1] considered the one-dimensional metric space given by an interval $\mathcal{R} = [-L, L) \subset \mathbb{R}$, $L > 0$, with the torus metric, i.e. the distance between two points $x, y \in \mathcal{R}$ is given by

$$d(x, y) = ||x - y|| = \min\{|x - y|, 2L - |x - y|\},$$

where $|x|$ is the absolute value of $x \in \mathbb{R}$. Furthermore, the analysis in [1] assumes that $\mathcal{A}_d = \mathcal{A}_s = \mathcal{R}$, i.e. for each content type $x \in \mathcal{R}$ there exists an agent in $\mathcal{A}_d$ who is most interested in content of type x, and there exists an agent in $\mathcal{A}_s$ who is most adept at producing content of type x.

In addition, the analysis in [1] considers a particular family $\mathcal{C}(L_C)$, $L_C > 0$, of community structures, given as follows. Let $N \geq 2$ be a given integer, and let

$$L_C = \frac{L}{N}, \tag{3}$$

where L is the half-length of the metric space $\mathcal{R} = [-L, L)$. Furthermore, let $\{m_k\}_{k=1}^N$ be a set of N evenly spaced points on the metric space $\mathcal{R} = [-L, L)$ given by

$$m_{k+1} = m_1 + 2L_C k, \qquad k = 1, \ldots, N - 1. \tag{4}$$

The set $\mathcal{C} = \{C^k = (C_d^k, C_s^k)\}_{k=1}^N$ of communities in the community structure $\mathcal{C}(L_C)$ is then given by N communities $C^k = (C_d^k, C_s^k)$, and for each community C^k the set of content consuming agents C_d^k, and the set of content producing agents C_s^k, are given by the intervals

$$C_d^k = [m_k - L_C, m_k + L_C) \text{ and } C_s^k = [m_k - L_C, m_k + L_C).$$

Furthermore, the allocations $\{\alpha_C(y)\}_{y \in \mathcal{R}}$ and $\{\beta_C(\cdot|y)\}_{y \in \mathcal{R}}$ of the community structure are given by

$$\alpha_{C^k}(y) = \begin{cases} E_p & y \in C_d^k \\ 0 & \text{otherwise} \end{cases}, \qquad k = 1, \ldots, N, \tag{5}$$

and

$$\beta_{C^k}(\cdot|y) = \begin{cases} E_q\delta(x - x_y^*) & y \in C_s^k \\ 0 & \text{otherwise} \end{cases}, \qquad k = 1, \ldots, N, \qquad (6)$$

where

$$x_y^* = \arg\max_{x \in \mathcal{R}} q(x|y)P_{C^k}(x).$$

The analysis in [1] shows that (under certain assumptions about the functions f and g that are used in Eqs. (1) and (2)) there always exists a covering Nash equilibrium within the family $\mathcal{C}(L_C)$, $L_C > 0$, of community structures.

3   Community Structure $\mathcal{C}(L_C, l_d)$

In this section we consider a family of community structures that is more general than the family $\mathcal{C}(L_C)$, $L_C > 0$, of the previous section, and study whether there exists a Nash equilibrium within this family.

More precisely, we consider the following family $\mathcal{C}(L_C, l_d)$ of community structures. Let $N \geq 2$ be a given integer and let $L_C = \frac{L}{N}$ as given by Eq. (3). In addition, let $\{m_k\}_{k=1}^N$ be a set of N evenly spaced points on the metric space $\mathcal{R} = [-L, L)$ as given by Eq. (4).

Given L_C, l_d, and m_k, $k = 1, \ldots, N$, as defined above, the set of communities $\mathcal{C} = \{C^k = (C_d^k, C_s^k)\}_{k=1}^N$ of the structure $\mathcal{C}(L_C, l_d)$ is then given by the intervals

$$C_d^k = [m_k - l_d, m_k + l_d) \quad \text{and} \quad C_s^k = [m_k - L_C, m_k + L_C).$$

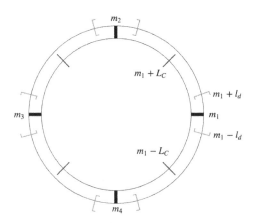

Fig. 1. The communities $\mathcal{C}$ for the case where $N = 4$ are illustrated. The metric space $\mathcal{R} = [-L, L)$ is shown as a ring to represent the torus (ring) metric. More precisely there are two rings: the outer ring represents the set of the content consumers $\mathcal{A}_d$, and the inner ring represents the set of content producers $\mathcal{A}_s$. The brackets on the outer ring bound the four consumption intervals C_d^k, $k = 1, \ldots, 4$, and the lines on the inner ring bound the four production intervals C_s^k, $k = 1, \ldots, 4$.

Figure 1 provides an illustration of these communities for the case of $N = 4$ communities. Furthermore, the allocations $\{\alpha_C(y)\}_{y \in \mathcal{R}}$ and $\{\beta_C(\cdot|y)\}_{y \in \mathcal{R}}$ of the community structure $\mathcal{C}(L_C, l_d)$ are as given by Eq. (5), and Eq. (6), respectively.

Note that for $l_d = L_C$, the community structure $\mathcal{C}(L_C, l_d) = \mathcal{C}(L_C, L_C)$ is identical to the community structure $\mathcal{C}(L_C)$ of the previous section that was analyzed in [1]. In particular, in this case the community structure $\mathcal{C}(L_C, L_C)$ is again a covering community structure, i.e. all agents belong to at least one community in $\mathcal{C}(L_C, L_C)$. As a result, we will focus on community structures $\mathcal{C}(L_C, l_d)$ where we have that $l_d < L_C$. In this case the community structure $\mathcal{C}(L_C, l_d)$, $0 < l_d < L_C$, is no longer a covering community structure. In particular, the content consuming agents in the sets

$$D^k = [m_k + l_d, m_{k+1} - l_d), \qquad k = 1, \ldots, N - 1,$$

and

$$D^N = [m_N + l_d, m_1 - l_d)$$

do not belong to any communities in $\mathcal{C}(L_C, l_d)$. On the other hand, note that all content producing agents $y \in \mathcal{R}$ do belong to at least one community C^k in the community structure $\mathcal{C}(L_C, l_d)$. In this sense, studying the existence of a Nash equilibrium within the family of community structures $\mathcal{C}(L_C, l_d)$ is studying whether there exists a Nash equilibrium from which some content consuming agents are excluded. We discuss the implications of such a Nash equilibrium in more detail in Sect. 5.

To study whether there exists a Nash equilibrium within the family $\mathcal{C}(L_C, l_d)$ of community structures as defined above, we use the following definitions. Let

$$x_y^*(l_d) = \arg\max_{x \in \mathcal{R}} q(x|y) \int_{-l_d}^{l_d} p(x|z)dz, \qquad y \in \mathcal{R}.$$

Furthermore, let the functions $G(y|L_C, l_d)$ and $H(y|L_C, l_d)$ be given by

$$G(y|L_C, l_d) = E_p E_q \int_{z=-L_C}^{L_C} p(x_z^*(l_d)|y)q(x_z^*(l_d)|z)dz - 2E_p E_q L_C c, \qquad y \in \mathcal{R},$$

and

$$H(y|L_C, l_d) = E_p E_q q(x_y^*(l_d)|y) \int_{z=-l_d}^{l_d} p(x_y^*(l_d)|z)dz - 2E_p E_q l_d c, \qquad y \in \mathcal{R},$$

where $c > 0$ is the cost for consuming a single content item.

In addition, we make the following assumptions about the functions f and g that are used in Eqs. (1) and (2).

Assumption 1. *The function $f : [0, \infty) \mapsto [0, 1]$ is given by*

$$f(x) = \max\{0, f_0 - ax\},$$

where $f_0 \in (0, 1]$ and $a > 0$. The function $g : [0, \infty) \mapsto [0, 1]$ is given by

$$g(x) = g_0,$$

where $g_0 \in (0, 1]$. Furthermore, we have that

$$f_0 g_0 > c. \qquad (7)$$

We note that the condition given by Eq. (7) is a necessary condition for a Nash equilibrium to exist, i.e. it is shown in [1] that if this condition is not true, then there does not exist a Nash equilibrium.

4   Main Results

In this section we present the main results of our analysis. Due to space constraints, we state the results without proofs, the proofs can be found in [4]. We first provide necessary and sufficient conditions for a community structure $\mathcal{C}(L_C, l_d)$ to be a Nash equilibrium.

Proposition 1. *Let the functions f and g be as given in Assumption 1. Furthermore, let L_C^* and l_d^* be such that*

$$0 < l_d^* < L_C^*, \quad \text{and} \quad L_C^* = \frac{L}{N},$$

where L is the half-length of the metric space $\mathcal{R} = [-L, L)$ and $N \geq 2$ is an integer. Then the community structure $\mathcal{C}(L_C^*, l_d^*)$ is a Nash equilibrium if, and only if, we have that

$$G(l_d^* | L_C^*, l_d^*) = 0, \quad \text{and} \quad H(L_C^* | L_C^*, l_d^*) \geq 0.$$

Our next result shows that there always exists a Nash equilibrium given that the half-length L of the metric space $\mathcal{R} = [-L, L)$ is large enough.

Proposition 2. *Let the functions f and g be as given in Assumption 1. If we have that*

$$L > 2 \left[\frac{f_0}{a} - \frac{c}{a g_0} \right],$$

then there always exists a community structure $\mathcal{C}(L_C, l_d)$, $0 < l_d < L_C$, that is a Nash equilibrium.

Proposition 2 states that for functions f and g as given in Assumption 1, there always exists a Nash equilibrium in the family of community structures $\mathcal{C}(L_C, l_d)$ given that L is large enough, i.e. if we have that $L > 2 \left[\frac{f_0}{a} - \frac{c}{a g_0} \right]$.

The next result provides a complete characterization of the values of L_C and l_d, $0 < l_d < L_C$, for which there exists a Nash equilibrium.

Proposition 3. *Let the functions f and g be as given in Assumption 1. Then the community structure $\mathcal{C}(L_C^*, l_d^*)$ with*

$$0 < l_d^* < L_C^*, \quad \text{and} \quad L_C^* = \frac{L}{N}$$

where $N \geq 2$ is an integer, is a Nash equilibrium if, and only if,

$$l_d^* = \frac{f_0}{a} - \frac{c}{ag_0}.$$

Note that the above result provides a complete characterization of the Nash equilibria within the family of community structures $\mathcal{C}(L_C, l_d)$. We discuss the interpretation of this result in more detail in the next section.

5   Conclusions

In this paper we show that there exists an additional family of Nash equilibria to the one identified in [1]. The Nash equilibria that we obtained have the property that some agents are excluded from the community structure, i.e. they do not belong to any of the communities. The reason for this is that these agents would have a negative utility in all of the communities that exist in the Nash equilibrium (see [4] for a formal derivation of this result). This means that these agents have the choice to either join a community where their utility would be negative, or not join any community at all (and obtain a utility of zero). Since in this situation agents are better off not joining any community, they are "marginalized". This outcome may come at a significant "social" cost to these agents. Studying this issue in depth is outside of the scope of this paper, but this is important and interesting future research. In particular, a natural question to ask in this context is whether, and how likely it is that the Nash equilibria that "marginalize" agents will indeed arise in information networks. This question can be studied formally by using the model in [1] to analyze the dynamics of community formation in information networks, and how the resulting dynamics can lead to the Nash equilibria that "marginalize" some agents.

References

1. Marbach, P.: Modeling and analysis of information communities. arXiv:1511.08904
2. Zadeh, R.B., Goel, A., Munagala, K., Sharma, A.: On the precision of social and information networks. In: Proceedings of the First ACM Conference on Online Social Networks, pp. 63–74. ACM (2013)
3. Hegde, N., Massoulié, L., Viennot, L.: Self-organizing flows in social networks. In: Moscibroda, T., Rescigno, A.A. (eds.) SIROCCO 2013. LNCS, vol. 8179, pp. 116–128. Springer, Cham (2013). https://doi.org/10.1007/978-3-319-03578-9_10
4. Carrington, M., Marbach, P.: Community structures in information networks. Technical report. http://www.cs.toronto.edu/~marbach/PUBL/gamenets_2019.pdf

Bargaining in Networks
with Socially-Aware Agents

Konstantinos Georgiou and Somnath Kundu[✉]

Department of Mathematics, Ryerson University, Toronto, ON M5B 2K3, Canada
{konstantinos,somnath.kundu}@ryerson.ca

Abstract. We introduce and characterize new stability notions in bargaining games over networks. Similar results were already known for networks induced by simple graphs, and for bargaining games whose underlying combinatorial optimization problems are packing-type. Our results are threefold. First, we study bargaining games whose underlying combinatorial optimization problems are *covering-type*. Second, we extend the study of stability notions when the networks are induced by *hypergraphs*, and we further extend the results to fully *weighted* instances where the objects that are negotiated have non-uniform value among the agents. Third, we introduce and characterize *new stability notions* that are naturally derived by polyhedral combinatorics and duality theory for Linear Programming. Interestingly, these new stability notions admit intuitive interpretations touching on *socially-aware* agents. Overall, our contributions are meant to identify natural and desirable bargaining outcomes as well as to characterize powerful positions in bargaining networks.

Keywords: Bargaining · Stable outcomes · Hypergraphs ·
Linear Programming

1 Introduction

Consider a set of agents, each of them demanding to receive a certain amount of service which can be offered by a number of available service providers. Choosing a specific service provider incurs some publicly known cost and serves a specific subset of the agents, possibly incurring different satisfaction to each of them. How would agents negotiate the cost distribution among them so as to agree on a global solution satisfying the demands of every player? Are there specific outcomes in which the cost of the services is fully covered by the agents, as well as agents' payment contributions are considered "fair"? We model this question as a General Covering Bargaining Game, and we characterize the existence of natural "fair" (or stable) outcomes. Our results are extensions to known stability notions for "Packing-Type Bargaining Games" in which the underlying graphs are simple, and agents treat all services (contracts) uniformly. Our findings further allow us to introduce and characterize new natural and relaxed notions of

K. Georgiou—Research supported in part by NSERC.

K. Avrachenkov et al. (Eds.): GameNets 2019, LNICST 277, pp. 128–150, 2019.
https://doi.org/10.1007/978-3-030-16989-3_10

stability (fair outcomes), whose interpretation is associated with socially-aware agents.

1.1 Related Work

Bargaining in networks has been studied extensively and for a long time, both in economics (as *cooperative games*) and sociology (as *network exchange theory*). The focus in economics has been the study of resource distributions (e.g. for two-sided markets [27, 28]), while in sociology the objective has been to understand the behaviour of agents who interact aiming to form relations of mutual benefit. For example, consider a primitive model, in which two players negotiate as to how to share the profit, say of 1, of a commonly required service. Given that each of them has *outside option* α and β, the so-called Nash bargaining solution [25] introduces the notion of a "fair" outcome in which each player receives her outside option and the surplus $1 - \alpha - \beta$ is divided evenly between the players.

As in the example above, the study of bargaining games entails the refinement of solution concepts with respect to notions of "fairness". Two such key notions are that of *stability* and *balance*. At a high level, an outcome is stable if the utility of every agent is at least as good as her outside option, i.e. the best utility an agent could have by deviating from a current agreement, say with another player, and by forming a new agreement. Balanced solutions were first introduced in [11, 27], and are a generalization of the Nash bargaining solution to networks. Interestingly, balanced solutions have been shown to agree with experimental results [30], however the focus of the current work is only on stable outcomes.

The framework of network bargaining games that we use in this work was first introduced by Kleinberg and Tardos in [22]. Their focus was a basic packing-type problem (matching), in which each agent could form up to one contract with a neighbour over a network (induced by a simple graph). Among others, Kleinberg and Tardos showed that such games have balanced solutions whenever they have stable solutions, and that (as we do in this work) the existence of stable solution is characterized by the integrality of a basic linear program relaxation for the associated combinatorial optimization problem (i.e. if a linear program has no discrepancy when compared to the exact but primitive integer program for the problem). Later, Bateni et al. [2] extended the work of [22] to bipartite (still simple) graphs in which some agents can engage in more than one contracts. Moreover, they showed that stable outcomes correspond to allocations in the *core* of the underlying coalition game, exhibiting this was a link between network bargaining (in matching and assignment games, previously studied in [8, 12, 13, 18, 28]) and cooperative game theory. More recently, Farczadi et al. [15] extended the results of the Kleinberg-Tardos model for networks with agents with general capacities, see also [19]. Relevant to our work is also [16] which considered again packing-type (matching) problems in which agents can bargain over a network from distance.

Variations of bargaining games have been studied extensively over the last decade. In [21] authors studied network bargaining games with general capacities.

Local dynamics in network bargaining games have been considered in [1,5,14]. [6] and [7] considered packing-type bargaining games with no capacity constraints, but with agents' utilities being nonlinear. [3] considered a local dynamical model of a one-sided exchange network (market) with transferable utilities and studied the dynamics of bargaining in such a market. In [9], the authors introduced alternative models for network bargaining based on instances with no stable outcomes, and in which players are both the negotiators and the negotiated objects. Finally, when bargaining instances do not admit stable solutions, the problem of minimally modifying the graph so as to inject stability was studied in [4,20].

1.2 Our Contributions and Paper Organization

In this paper we follow the techniques and further generalize part of the work of [2,15,22]. The common underlying bargaining games in these results pertained to a specific packing-type problem (matching), defined over simple graphs (players interactions were only binary), and where the subject of bargaining had a uniform value for all agents (contracts were worth the same to all agents). In contrast, we study covering-type games, where agents are competing to receive services, and we provide, as in the previous papers, a characterization of the existence of *stable* solutions. Our model is more general, in that we allow that services are of non-uniform value, i.e. the same service may provide different satisfaction to each agent. More importantly, our bargaining games are defined over networks induced by *hypergraphs*, i.e. services (or contracts in the previous packing-type problems) are not binary relations. As a consequence, we also introduce natural families of *relaxed notions of stability* based on cutting planes for the linear program formulations of the underlying combinatorial optimization problems. Interestingly, these new notions of stability admit an intuitive interpretation pertaining to *socially-aware* agents. Our findings also find applications to bargaining games whose combinatorial optimization problems do not admit IP formulations where constraints are associated only with agents. In particular, our relaxed stability notions apply also when not all vertices of a network are agents, rather they are present only to facilitate agreements. Notable, none of the previously known results were able to address stability in such networks.

In Sect. 2 we introduce the general covering-type problems we study in this work. In Sect. 2.1 we provide the game-theoretic perspective of these problems and we define the standard notions of feasibility and stability in bargaining outcomes. Then in Sect. 2.2 we give the combinatorial perspective of the covering-type problems, as well as we review the tools from Linear Programming that are used extensively in our work. Section 3 is devoted to the characterization of the existence of stable solutions. In particular, one of our main contributions, Theorem 1 is proven in Sects. 3.1 and 3.2. Then, in Sect. 4 we study relaxed notions of stability. In Sect. 4.1 we introduce the key concept of critical constraints upon which we will rely to relax stability. Section 4.2 introduces new notions of stability, and finally in Sect. 4.3 we provide a characterization of their existence, as well as we give a natural interpretation for them.

2 Covering-Type Problems in Hypergraphs; Bargaining Games vs Combinatorial Optimization

The purpose of this section is to formally introduce bargaining games in hypergraphs and their counterparts, combinatorial optimization problems. Common to both is the underlying input, which formally speaking consists of a hypergraph $\mathcal{H} = (V, E)$, where $e \subseteq V$ for each $e \in E$. Set V will be called the set of agents (or players), and E will be referred to as the set of services. $\mathbb{Q}_{++}$ below denotes the set of positive rational numbers. We consider functions $d : V \mapsto \mathbb{Q}_{++}, c : E \mapsto \mathbb{Q}_{++}, \rho : V \times E \mapsto \mathbb{Q}_{++}$. For each $i \in V, e \in E$ we will commonly write $d_i, c_e, \rho_{i,e}$, instead of $d(i), c(e), \rho(i, e)$, respectively. Moreover, d_i will be called the demand of agent i, c_e will be called the cost of service e, and $\rho_{i,e}$ will be called the satisfaction of agent i for service e. Whenever $i \in e \in E$ we will say that service e covers i. For each $i \in V$ we denote by T_i the set of services covering i, i.e. $T_i = \{e \in E : i \in e\}$. Altogether, we will refer to the tuple $\mathcal{B} = (\mathcal{H} = (V, E), d, c, \rho)$ as a *covering-type problem*. In order to avoid degenerate cases, the silent assumption in all covering-type problems is that for each $i \in V$, $\sum_{e \in T_i} \rho_{i,e} \geq d_i$. In other words, all services are enough to meet all players' demands.

Next we view covering-type problems $\mathcal{B}$ under two different lenses, one game-theoretic and one combinatorial. At a high level, the game-theoretic problem will attempt to understand $\mathcal{B}$ from the perspective of selfish and rational players, set V, who are willing to cover (part of) the expenses for choosing enough many services (set E) so as to cover their needs/demands (values $d_i, i \in V$). The combinatorial lens will view $\mathcal{B}$ from the perspective of a central authority who is attempting to choose the least expensive set of contracts so as to satisfy all demands. Later on, we bridge the two perspectives by identifying polyhedral combinatorial properties of the combinatorial problem that characterize when "stable" solutions to the game-theoretic bargaining problem exist.

2.1 Cost Sharing in Bargaining Games over Networks

The purpose of this section is to introduce covering-type bargaining games in networks induced by hypergraphs, along with their solution concepts. Formally, a *general covering bargaining game* (or simply, bargaining game) is given by covering-type problem $\mathcal{B} = (\mathcal{H} = (V, E), d, c, \rho)$.

Bargaining game $\mathcal{B}$ corresponds to a set of players V, each of them i looking to be serviced by services that provide at least d_i satisfaction (in the uniform case, where all satisfactions are 1, d_i is the number of services requested). Naturally, the set of agents, together with services induce a hypergraph $\mathcal{H}$, in which hyperedges $e \subseteq V$ are identified by the subset of players they serve. Given that service e incurs cost $c_e > 0$, we would like to understand the bargaining dynamics in the induced network, when it comes to choosing a collection of services and paying for them. The underlying assumption, as in any game-theoretic problem, is that agents are selfish and rational. In our case, agents cannot compromise on the number of the services each receives (indicated by their demands). As all agents will need to naturally cover the entire cost of the chosen services, each agent would like to minimize her contribution toward purchasing (choosing) any of the services. In this direction, we introduce the notion of a bargaining outcome, which will be central in identifying the bargaining dynamics.

Definition 1 (Bargaining Feasible Outcome). *A bargaining feasible outcome (or simply an outcome) of bargaining game $\mathcal{B}$ is a tuple $\mathcal{F} = (A \subseteq E, \{P_{i,e}\}_{i \in V, e \in E})$, with $P_{i,e} \in \mathbb{R}_+$, satisfying the following properties:*

- *(Demand Satisfaction) For every $i \in V$, $\sum_{e \in T_i \cap A} \rho_{i,e} \geq d_i$.*
- *(Cost Recovery) For every $e \in A$, $\sum_{i \in V} P_{i,e} = c_e$.*

Bargaining feasible outcome $\mathcal{F}$ of bargaining game $\mathcal{B}$ specifies exactly a subset A of the services that are chosen to meet the demands of all players. Intuitively, every player i needs to contribute some non-negative payments $P_{i,e}$ for each chosen service, and these payments should cover its cost. Note that we allow positive contributions $P_{i,e}$ for players i even when they are not covered by a service e (this will become relevant when we will introduce relaxed notions of stability).

Given outcome $\mathcal{F}$ of a bargaining game, we partition the set of agents in two disjoint sets; the set V_* of *oversaturated* agents i for which $\sum_{e \in T_i \cap A} \rho_{i,e} > d_i$, and the set of *tight* agents $V \setminus V_*$. Note that for each tight agent i, and by the definition of feasible outcomes, we have $\sum_{e \in T_i \cap A} \rho_{i,e} = d_i$, i.e. agent i meets her demand exactly, while oversaturated agents receive strictly more service satisfaction than their demands.

As our goal is to identify "desirable" outcomes as well as the powerful positions in a bargaining network, the notion of feasible outcomes can be refined as follows.

Definition 2 (Stable Outcome). *A bargaining feasible outcome $\mathcal{F}$ of bargaining game $\mathcal{B}$ is called stable if the following properties are satisfied:*

- *(Greed) $P_{i,e} > 0$, implies that agent i is tight, service e is chosen and $i \in e$, i.e. $i \in V \setminus V_*$ and $i \in e \in A$.*
- *(Envy-Free) For every $f = \{i_1, \ldots, i_l\} \notin A$, and for every $e_j \in A \cap T_{i_j}, j = 1, \ldots, l$,*

$$\sum_{j=1}^{l} \frac{\rho_{i_j,f}}{\rho_{i_j,e_j}} P_{i_j,e_j} \leq c_f. \tag{1}$$

Stable outcomes are meant to propose a refinement of feasible outcomes that are desirable by the agents (meaning that a bargaining process may converge to such an outcome) by requiring fair payments. Indeed, an agent should never pay for a service that does not serve her, or a service that is not chosen. Every service $e \in E$ can be thought as a potential coalition among $e \subseteq V$ who choose to pay for service $e \in A$, and hence $f \notin A$ can be thought as coalitions that are not formed. A positive payment may be required only by tight agents, as otherwise an oversaturated agent might feel in a powerful negotiation power when a solution is proposed that oversatisfies her demands. Finally, for any unformed coalition $f \notin A$ (i.e. a service not in the solution), and every agent $i \in f$ consider the maximum payment P_{i,e_i} that i makes over all services in A. For the sake of the argument, assume that satisfaction rates are uniform. If it happened that $\sum_{i \in f} P_{i,e_i} > c_f$, then each of the agents i would like to leave coalition e_i so that all of them form coalition f, effectively sharing the cost c_f and reducing their previous maximum payments. The reader may think of the latter requirement as the standard stability notion of coalition games with transferable payoffs. Now, if the satisfaction rates are not uniform, as it is the case in our model, then the same argument holds for normalized payments $\rho_{i_j,f} \cdot P_{i_j,e_j}/\rho_{i_j,e_j}$. The latest expression is interesting in its own right, as it provides a form of satisfaction conversion between services f, e_j using the relative payment per unit of satisfaction of player i_j for service e_j. Finally, it is important to notice that expression (1) can be used to introduce (and hence generalize existing) notions of agents' *outside options* for our covering-type games.

2.2   The Underlying Covering-Type Combinatorial Optimization Problem

Now we turn our attention to the underlying combinatorial perspective of a covering-type problem $\mathcal{B} = (\mathcal{H} = (V, E), d, c, \rho)$. We interpret $\mathcal{B}$ as a combinatorial optimization problem, in which a central authority is trying to find a feasible collection of services $A \subseteq E$ so as to satisfy all demands d_i of each agent i. Moreover, among the set of feasible collection of services, one would like to identify the least costly, i.e. to minimize the sum of costs c_e for services $e \in A$.

Combinatorial problem $\mathcal{B}$ admits a natural formulation as an Integer Program (IP). We introduce an indicator, 0-1, variable x_e for every $e \in E$ which is thought as 1 if and only if e is chosen in a feasible solution. Requiring that each agent i receives at least as many services as her demand d_i (with respect to her satisfactions) and minimizing the overall induced cost, we obtain the following exact formulation of $\mathcal{B}$.

$$
\begin{aligned}
\min \quad & \sum_{e \in E} c_e\, x_e && (F_{IP}(\mathcal{B})) \\
\text{s.t.} \quad & \sum_{e \in T_i} \rho_{i,e} x_e \geq d_i, && \forall i \in V \\
& -\mathbf{x} \geq -\mathbf{1}, \\
& \mathbf{x} \geq \mathbf{0}, \\
& \mathbf{x} \in \mathbb{Z}^{|E|}
\end{aligned}
$$

In what follows, we refer to this formulation as F_{IP}. Given as input $\mathcal{B}$, we denote by $optF_{IP}(\mathcal{B})$ its optimal value (note that the IP is feasible and bounded with rational coefficients, and hence it always admits and optimal solution). Next we overview some standard tools from combinatorial optimization that will be useful later on. First we introduce the so-called linear program (LP) relaxation of the IP above, which is obtained by dropping the integrality condition $\mathbf{x} \in \mathbb{Z}^{|E|}$. We will denote the resulting LP by F_{LP}. Given input covering-type problem $\mathcal{B}$, we denote by $optF_{LP}(\mathcal{B})$ its optimal solution (the LP is feasible and bounded, hence it always attains an optimal solution). By definition, and for every covering-type problem $\mathcal{B}$, we have that $optF_{LP}(\mathcal{B}) \leq optF_{IP}(\mathcal{B})$. In particular, the ratio $\frac{optF_{LP}(\mathcal{B})}{optF_{IP}(\mathcal{B})}$ is known as the *integrality gap* of F_{LP} on input $\mathcal{B}$ and measures the discrepancy between F_{LP} and F_{IP} for the same instance. The notion of the so-called *integrality gap* of the LP relaxation measures the worst case discrepancy between an IP and its LP relaxation, over all instances;

$$\inf_{\mathcal{B}} \frac{optF_{LP}(\mathcal{B})}{optF_{IP}(\mathcal{B})}.$$

It follows that the integrality gap of F_{LP} on input $\mathcal{B}$ is at most 1, and it is equal to 1 if and only if there exists an integral optimal solution to F_{LP} with input $\mathcal{B}$. Similarly, the integrality gap of F_{LP} is 1 if and only if F_{LP} admits an integral optimal solution for every input $\mathcal{B}$.

Example 1 (Vertex Cover). In VERTEX-COVER one is given a simple graph $G = (V_0, E_0)$. Feasible solutions are subsets of the vertices $S \subseteq V_0$ with the property that for every edge $e = \{i, j\}$, at least one of the endpoints i, j lies in S (the chosen subset of the vertices is called a vertex cover). Let $\mathcal{H}$ denote the line graph of G. Then, *Vertex-Cover* is the covering-type problem $\mathcal{B} = (\mathcal{H}, d, c, \rho)$, where $d_e = 1$ for all $e \in E_0$, $c_i = 1$ for all $i \in V_0$, and $\rho_{e,i} = 1$ for all $i \in V_0$ and $e \in E_0$. Note that the set of agents are the edges E_0 and the set of vertices V_0 forms the set of services. Each agent $e = \{i, j\}$ needs at least one service chosen among i, j. The LP relaxation of the exact IP formulation of the problem reads as

$$\min \quad \sum_{i \in V_0} x_i \qquad\qquad\qquad\qquad (2)$$
$$\text{s.t.} \quad x_i + x_j \geq 1, \qquad\qquad \forall \{i, j\} \in E_0$$
$$0 \leq \mathbf{x} \leq 1.$$

It is well known that the integrality gap of the above LP is 1 if G is bipartite, while in general the integrality gap can be as small as $1/2$.

3   Characterization of Stable Outcomes in Hypergraphs

In this section we characterize the existence of stable solutions of general bargaining games $\mathcal{B}$. Note that our result is an extension of known results when the underlying network of $\mathcal{B}$ is a simple graph, and hence every service can potentially serve exactly two agents. The extension to hypergraphs will be essential when we will introduce natural relaxed notions of stability in Sect. 4.

Theorem 1. *Bargaining game $\mathcal{B}$ admits a stable outcome if and only if the integrality gap of $F_{LP}(\mathcal{B})$ is 1.*

We comment throughout the exposition of our proof how one needs to extend/modify existing arguments (for simple graphs) to obtain our results for hypergraphs and for non-uniform satisfactions and demands. However, the key ingredient to establishing Theorem 1 is actually the introduction of a proper definition of stability, which is also tailored to the notion of over-saturation. In fact, our generalized/modified notions of stability and over-saturation simplify to the already studied notions of stability and over-saturation when all demands and satisfactions are 1 and all contracts are of size exactly 2. More specifically, our main goal was to establish a characterization of the existence of stable solutions for our generalized bargaining instances based on the integrality of natural LP relaxations. As such, the new proposed notion of stability (the one of Theorem 1) was derived mechanically, as it is tailored to the integrality of a certain LP. Whether the associated notions of over-saturation that one needs to consider is practical or not is outside the scope of this paper, nevertheless it is not difficult to argue, at a high level, why it is natural.

It is important to note that Theorem 1 is tailored to the exact formulation F_{IP}. Indeed, Theorem 1 characterizes the existence of stable solutions to the bargaining when the set of feasible solutions to the combinatorial optimization problem can be described by one linear constraint associated with each agent, i.e. when F_{IP} is an exact formulation to the covering-type problem. In particular, if the set of feasible solutions to the combinatorial optimization problem requires additional constraints in order to be determined, or if "redundant constraints" are added to the IP formulation (not necessarily redundant for the LP relaxation), then Theorem 1 does not apply. In other words, the existence of "natural" solution concepts (that of stable solutions as per Definition 2) to bargaining games are derived *mechanically* by structural properties in polyhedral combinatorics, at least in special cases.

Example 2. Consider the bargaining game induced by VERTEX-COVER of Example 1, for some input graph $G = (V_0, E_0)$. If G is bipartite, then the bargaining game admits a stable outcome. On the other hand, consider the simple graph

$$G_0 = (\{1, 2, 3\}, \{\{1, 2\}, \{2, 3\}, \{1, 3\}\}).$$

Setting $x_1 = x_2 = x_3 = 1/2$ is feasible to LP (2), hence the optimal value to the LP is at most $3/2$. At the same time, no less than 2 services among $\{1, 2, 3\}$ need to be chosen in any solution to the covering-type problem (independently of its cost). The integrality gap of the LP is not 1, hence, according to Theorem 1 the bargaining game admits no stable solution.

The driving force behind proving Theorem 1 is duality theory and complementary slackness conditions. First we propose the dual linear program of $F_{LP}(\mathcal{B})$, where $\mathcal{B} = (\mathcal{H} = (V, E), d, c, \rho)$, which reads as follows;

$$\max \quad \sum_{i \in V} d_i \, y_i - \sum_{e \in E} u_e \qquad \qquad (F_{LP}^D(\mathcal{B}))$$

$$\text{s.t.} \quad \sum_{i \in e} \rho_{i,e} y_i - u_e \leq c_e, \qquad \forall e \in E$$

$$\mathbf{y}, \mathbf{u} \geq \mathbf{0},$$

We denote by $optF_{LP}^D(\mathcal{B})$ the optimal solution to the above LP.

Now consider a primal dual pair of feasible solutions $\bar{\mathbf{x}}, (\bar{\mathbf{y}}, \bar{\mathbf{u}})$ to $F_{LP}(\mathcal{B})$ and $F_{LP}^D(\mathcal{B})$, respectively. Since $F_{LP}(\mathcal{B})$ admits an optimal solution for every $\mathcal{B}$ such a pair always exists (by strong duality), and they are each optimal to the primal and dual LPs if and only if the so-called *complementary slackness conditions* hold true;

$$\left(\sum_{e \in T_i} \bar{\rho}_{i,e} \bar{x}_e - d_i \right) \bar{y}_i = 0, \ \forall i \in V \qquad (3)$$

$$(\bar{x}_e - 1) \, \bar{u}_e = 0, \ \forall e \in E \qquad (4)$$

$$\left(\sum_{i \in e} \rho_{i,e} \bar{y}_i - \bar{u}_e - c_e \right) \bar{x}_e = 0, \ \forall e \in E \qquad (5)$$

3.1   Integrality from Stability

In this section we prove the "only if" claim of Theorem 1, that is we prove the statement below, whose proof follows closely known arguments for simple graphs (one needs to only normalize payments with respect to satisfaction rates).

Lemma 1. *If bargaining game $\mathcal{B}$ admits a stable outcome then the integrality gap of $F_{LP}(\mathcal{B})$ is 1.*

So, fix some bargaining game $\mathcal{B} = (\mathcal{H} = (V, E), d, c, \rho)$ and a stable (and feasible) outcome $\mathcal{F} = (A \subseteq E, \{P_{i,e}\}_{i \in V, e \in E})$. We prove Lemma 1 by finding an *optimal solution* to $F_{LP}(\mathcal{B})$ which is also integral.

We define the following primal-dual pair $\bar{\mathbf{x}}, (\bar{\mathbf{y}}, \bar{\mathbf{u}})$ of feasible solutions (as we will shortly prove) to $F_{LP}(\mathcal{B})$ and $F_{LP}^D(\mathcal{B})$, respectively.

$$\bar{x}_e := \begin{cases} 1, \text{ if } e \in A \\ 0, \text{ if } e \notin A \end{cases} \qquad (6)$$

$$\bar{y}_i := \begin{cases} \max \left\{ \frac{P_{i,e}}{\rho_{i,e}} : \ e \in T_i \cap A \right\}, \text{ if } i \text{ is tight} \\ 0, \qquad\qquad\qquad\qquad\quad \text{ if } i \text{ is oversaturated} \end{cases} \qquad (7)$$

and

$$\bar{u}_e := \begin{cases} \sum_{i \in e} \rho_{i,e} \, \bar{y}_i - c_e, \text{ if } e \in A \\ 0, \qquad\qquad\qquad \text{ if } e \notin A \end{cases} \qquad (8)$$

Lemma 2. $\bar{\mathbf{x}}, (\bar{\mathbf{y}}, \bar{\mathbf{u}})$ *are feasible to $F_{LP}(\mathcal{B})$ and $F_{LP}^D(\mathcal{B})$, respectively.*

Proof. $\mathcal{F} = (A \subseteq E, \{P_{i,e}\}_{i \in V, e \in E})$ is a feasible outcome of the bargaining game $\mathcal{B}$. As such, by Definition 1 we have $d_i \leq \sum_{e \in T_i \cap A} \rho_{i,e} = \sum_{e \in T_i} \rho_{i,e} \bar{x}_e$. Given also that $0 \leq \bar{\mathbf{x}} \leq 1$, we see that $\bar{\mathbf{x}}$ is feasible to $F_{LP}(\mathcal{B})$.

Now we show that $(\bar{\mathbf{y}}, \bar{\mathbf{u}})$ is feasible to $F_{LP}^D(\mathcal{B})$. First we argue that $\bar{\mathbf{y}}, \bar{\mathbf{u}}$ are non-negative vectors. First, $\bar{\mathbf{y}}$ follows by the non-negativity of payments $P_{i,e}$ (see Definition 1) and the fact that service satisfactions are strictly positive. Second, let $e \in E$ arbitrary. If $e \notin A$ we have $\bar{u}_e \geq 0$ by construction. Otherwise, consider some service $e \in A$. For every $i \in e$, denote by $f_i := \arg\max \left\{ \frac{P_{i,f}}{\rho_{i,f}} : f \in T_i \cap A \right\}$. Then,

$$\bar{u}_e = \sum_{i \in e} \rho_{i,e} \, \bar{y}_i - c_e = \sum_{i \in e} \rho_{i,e} \frac{P_{i,f_i}}{\rho_{i,f_i}} - c_e \geq \sum_{i \in e} \rho_{i,e} \frac{P_{i,e}}{\rho_{i,e}} - c_e = 0,$$

where the last equality follows from Definition 1 (Cost Recovery).

It remains to prove that for all $e \in E$ we have $\sum_{i : e \in T_i} \rho_{i,e} \bar{y}_i - \bar{u}_e \leq c_e$. We examine two cases. If $e \in A$, then note that by construction (see (7) and (8)), the constraint is satisfied tightly. Otherwise, if $e \notin A$, we rely on that the feasible outcome is stable, and indeed we have

$$\sum_{i : e \in T_i} \rho_{i,e} \bar{y}_i - \bar{u}_e \overset{(8)}{=} \sum_{i : e \in T_i} \rho_{i,e} \bar{y}_i \overset{(7)}{\leq} \sum_{i : e \in T_i} \rho_{i,e} \max_{f \in T_i \cap A} \left\{ \frac{P_{i,f}}{\rho_{i,f}} \right\} \overset{(1)}{\leq} c_e.$$

$\square$

By Lemma 2, it follows that if $\bar{\mathbf{x}}, (\bar{\mathbf{y}}, \bar{\mathbf{u}})$ satisfy complementary slackness conditions, then $\bar{x}$ is optimal to $F_{LP}(\mathcal{B})$. Since also $\bar{x}$ is integral, that would imply that the integrality gap of $F_{LP}(\mathcal{B})$ is 1. Therefore, Lemma 1 follows by the lemma below.

Lemma 3. $\bar{\mathbf{x}}, (\bar{\mathbf{y}}, \bar{\mathbf{u}})$ *satisfy complementary slackness conditions* (3), (4), (5).

Proof. First we examine (3) for an arbitrary agent i. If i is oversaturated, then by (7) we have $\bar{y}_i = 0$. If i is tight, then by the definition of tight agents, we have $d_i = \sum_{e \in T_i \cap A} \rho_{i,e} = \sum_{e \in T_i} \bar{\rho}_{i,e} \bar{x}_e$, as wanted.

Second, we study (4) for arbitrary $e \in E$. If $e \in A$, then by (6) we have $\bar{x}_e = 1$, while if $e \notin A$, then by (8) we have $\bar{u}_e = 0$. In any case, condition (4) is satisfied.

Third, we examine (5) for an arbitrary $e \in A$. If $e \notin A$, then by (6) we have $\bar{x}_e = 0$ and the condition is satisfied. Otherwise, $e \in A$. But then, note that by (8), $\bar{u}_e$ was chosen so as to make constraint $\sum_{i : e \in T_i} \rho_{i,e} \bar{y}_i - \bar{u}_e \leq c_e$ tight, independently of the valuations of $\bar{y}_i$ for agents $i \in e$. $\square$

3.2   Stability from Integrality

In this section we prove the "if" claim of Theorem 1, that is we prove that

Lemma 4. *If for some covering-type problem $\mathcal{B}$ the integrality gap of $F_{LP}(\mathcal{B})$ is 1, then the underlying bargaining game $\mathcal{B}$ admits a stable outcome.*

For a fixed covering-type problem $\mathcal{B}$, consider an optimal integral solution $\bar{\mathbf{x}}$ to $F_{LP}(\mathcal{B})$. Next we propose an outcome to the bargaining game $\mathcal{B}$, and subsequently we show it is feasible (as per Definition 1) and stable (as per Definition 2). For this, we set

$$A := \{e \in E : \bar{x}_e = 1\}. \tag{9}$$

Notice that by construction, $\bar{\mathbf{x}}$ is feasible to $F_{IP}(\mathcal{B})$, and hence for each $i \in V$ we have $\sum_{e \in T_i \cap A} \rho_{i,e} = \sum_{e \in T_i} \bar{\rho}_{i,e} \bar{x}_e \geq d_i$, hence each agent meets her demand as per the (partial) requirement of feasible outcomes. In other words, A is a feasible solution to the covering-type problem $\mathcal{B}$.

In order to propose payments for players, we need a couple of observations. These arguments (Lemma 2 and in particular Lemma 5 below) require a much more delicate treatment than in the case of simple graphs with uniform demands and satisfactions.

Observation 2. *For each $e \in A$, there is at least one agent $i \in A$ that is tight.*

Proof. Since $\bar{\mathbf{x}}$ is optimal to $F_{LP}(\mathcal{B})$, it is also optimal to $F_{IP}(\mathcal{B})$, hence A is an optimal solution to the covering-type problem $\mathcal{B}$. For the sake of contradiction, assume that there is some $e_0 \in A$ for which

$$\sum_{e \in T_i} \rho_{i,e} x_e > d_i,$$

for all $i \in e_0$. Recall that for all $e \in E$ we have $\rho_{i,e} \geq 0$ and $c_e > 0$. As a result, there exists a small enough $\epsilon > 0$ so that by updating $\bar{x}_{e_0} \leftarrow \bar{x}_{e_0} - \epsilon$, vector $\bar{x}$ remains feasible and has cost strictly less than $\sum_{e \in A} c_e = optF_{LP}(\mathcal{B})$, a contradiction to the optimality of $\bar{\mathbf{x}}$. $\qquad\square$

Next we recall that since $\bar{\mathbf{x}}$ is optimal to $F_{LP}^D(\mathcal{B})$ and by strong duality, $F_{LP}^D(\mathcal{B})$ admits an optimal solution, call it $(\bar{\mathbf{y}}, \bar{\mathbf{u}})$, and in particular, the primal-dual pair of feasible solutions satisfy complementary slackness conditions (3), (4), (5).

Lemma 5. *Given $(\bar{\mathbf{y}}, \bar{\mathbf{u}})$, and for every $e \in A$, there exists a distribution $\{\lambda_{i,e}\}_{i \in e}$ satisfying*

$$\rho_{i,e}\,\bar{y}_i - \lambda_{i,e}\bar{u}_e \geq 0, \forall i \in e, \tag{10}$$

and whose support lies only within the tight agents of e.

Proof. Consider arbitrary $e \in A$ (and hence $\bar{x}_e = 1$). In what follows we construct non-negative $\{\lambda_{i,e}\}_{i \in e}$ with

$$\sum_{i \in e} \lambda_{i,e} = 1 \tag{11}$$

$$\lambda_{i,e} = 0, \forall i \in e \cap V_* \tag{12}$$

satisfying (10).

Since $\bar{x}_e = 1$, by complementary slackness condition (5), we know that

$$\sum_{i \in e} \rho_{i,e} \bar{y}_i - \bar{u}_e = c_e, \tag{13}$$

where $c_e > 0$. Recall that by Observation 2, service e contains at least one tight agent. Hence, if $\bar{u}_e = 0$ we put all weight to that agent and we are done.

In what follows, we assume $\bar{u}_e > 0$, and denote by e_* the set of agents within e that are oversaturated (i.e. not tight), and note that $e \setminus e_* \neq \emptyset$ (by Observation 2). For all $i \in e_*$ we set $\lambda_{i,e} = 0$, as required by (12). By complementary slackness condition (3) we conclude that $\bar{y}_i = 0$ whenever $i \in e_*$. Hence, for all oversaturated agents i, we have that (10) is satisfied tightly. Moreover, (13) can be rewritten as

$$\sum_{i \in e \setminus e_*} (\rho_{i,e} \bar{y}_i - \lambda_{i,e} \bar{u}_e) = c_e,$$

for arbitrary $\lambda_{i,e}$ satisfying (11). It remains to prove that $\{\lambda_{i,e}\}_{i \in e \setminus e_*}$ can be indeed chosen to be non-negative so as to also satisfy $\rho_{i,e} \bar{y}_i - \lambda_{i,e} \bar{u}_e \geq 0$, for all $i \in e \setminus e_*$.

To that end, notice that since $c_e > 0$ and by (13), we have $\sum_{i \in e \setminus e_*} \rho_{i,e} \bar{y}_i > \bar{u}_e \geq 0$. Then, set

$$\lambda_{i,e} = \frac{\rho_{i,e} \bar{y}_i}{\sum_{j \in e \setminus e_*} \rho_{j,e} \bar{y}_j}.$$

But then,

$$\rho_{i,e} \bar{y}_i - \lambda_{i,e} \bar{u}_e = \rho_{i,e} \bar{y}_i \left(1 - \frac{\bar{u}_e}{\sum_{j \in e \setminus e_*} \rho_{j,e} \bar{y}_j} \right) > \rho_{i,e} \bar{y}_i \geq 0,$$

where the last inequality follows by dual feasibility and that $\rho_{i,e} \geq 0$. $\square$

Now, for any fixed collection of distributions $\{\{\lambda_{i,e}\}_{i \in e}\}_{e \in A}$ as per Lemma 5, we define payments

$$P_{i,e} := \begin{cases} \rho_{i,e} \bar{y}_i - \lambda_{i,e} u_e, & \text{if } e \in A \text{ and } i \in e \\ 0, & \text{otherwise} \end{cases} \tag{14}$$

Altogether, (9) and (14) above determine outcome $\mathcal{F} = (A \subseteq E, \{P_{i,e}\}_{i \in V, e \in E})$. The following two lemmata verify that $\mathcal{F}$ is a feasible and stable, implying Lemma 4.

Lemma 6. *Outcome* $\mathcal{F} = (A \subseteq E, \{P_{i,e}\}_{i \in V, e \in E})$ *is feasible (as per Definition 1).*

Proof. Set $A \subseteq E$, as defined in (9) was already shown to satisfy Demand Satisfaction. Now, Lemma 5 implies, first, that all payments $P_{i,e}$ are indeed non-negative. In order to show that they also satisfy Cost Recovery, consider arbitrary

$e \in A$, and not that by construction, only tight agents within e may have positive payments. Therefore, we have

$$\sum_{i \in V} P_{i,e} = \sum_{i \in e \backslash V_*} P_{i,e} \overset{(14)}{=} \sum_{i \in e \backslash V_*} (\rho_{i,e}\, \bar{y}_i - \lambda_{i,e} u_e) \overset{(Lemma\ 5)}{=} \sum_{i \in e \backslash V_*} \rho_{i,e}\, \bar{y}_i - u_e.$$

$$(15)$$

By complementary slackness condition (3), oversaturated agents i in e have $\bar{y}_i = 0$, and hence (15) further equals,

$$\sum_{i \in e} \rho_{i,e}\, \bar{y}_i - u_e = c_e$$

where the last equality is due to dual feasibility and complementary slackness condition (5), since $\bar{x}_e = 1$.   □

Lemma 7. *Outcome $\mathcal{F} = (A \subseteq E, \{P_{i,e}\}_{i \in V, e \in E})$ is stable (as per Definition 2).*

Proof. First we observe that by the definition of the payments (14), $P_{i,e} = 0$ if $i \notin e$ or if $e \notin A$. Now consider some $e \in A$ and some oversaturated agent $i \in e$. By Lemma 5 we have $\lambda_{i,e} = 0$, and by complementary slackness condition (3) we have $\bar{y}_i = 0$. Hence, by (14) we obtain $P_{i,e} = 0$, overall concluding property Greed.

Now we verify that $\mathcal{F}$ is also Envy-Free. For this, consider arbitrary $f = \{i_1, \ldots, i_l\} \notin A$, and arbitrary $e_j \in A \cap T_{i_j}$ for $j = 1, \ldots, l$. We have

$$\sum_{j \in f} \frac{\rho_{j,f}}{\rho_{j,e_j}} P_{j,e_j} = \sum_{j \in f \backslash V_*} \frac{\rho_{j,f}}{\rho_{j,e_j}} P_{j,e_j} \qquad \text{(by Greed)}$$

$$= \sum_{j \in f \backslash V_*} \frac{\rho_{j,f}}{\rho_{j,e_j}} \left(\rho_{j,e_j}\, \bar{y}_j - \lambda_{j,e_j} \bar{u}_{e_j} \right) \qquad \text{(by (14))}$$

$$\leq \sum_{j \in f \backslash V_*} \rho_{j,f}\, \bar{y}_j \qquad (\lambda_{j,e_j}, \bar{u}_{e_j} \geq 0)$$

$$= \sum_{j \in f \backslash V_*} \rho_{j,f} \bar{y}_j - \bar{u}_f \qquad (\bar{u}_f = 0, \text{since} f \notin A \text{ and by (4)})$$

$$= \sum_{j \in f} \rho_{j,f} \bar{y}_j - \bar{u}_f \qquad (\bar{y}_j = 0 \text{ for } j \in V_*, \text{ by (3)})$$

$$\leq c_f. \qquad ((\bar{\mathbf{y}}, \bar{\mathbf{u}}) \text{ feasible to } F^D_{LP}(\mathcal{B}))$$

□

4   Stability Notions Based on Socially-Aware Agents

4.1   Critical Constraints

Consider a covering-type problem $\mathcal{B} = (\mathcal{H} = (V, E), d, c, \rho)$, along with the underlying bargaining game. The following definition is the starting point toward deriving a new notion of stability.

Definition 3 (Critical Constraint). *Let $R \subseteq E$, $t \in \mathbb{Q}_{++}$ and $\sigma : R \mapsto \mathbb{Q}_{++}$. Triplet (R, t, σ) is called a critical constraint for $\mathcal{B}$ if for every $A \subseteq E$ satisfying demand satisfaction (as per Definition 1), we have*

$$\sum_{e \in A \cap R} \sigma_e \geq t.$$

A critical constraint for $\mathcal{B}$ models a property that any feasible solution to a bargaining game satisfies, at least when it comes to demand satisfaction. In the combinatorial optimization world, (R, t, σ) corresponds exactly to the *redundant constraint*

$$\sum_{e \in R} \sigma_e x_e \geq t. \tag{16}$$

of $F_{IP}(\mathcal{B})$, i.e. a constraint that is derivable by the remaining constraints of the integer program. In other words, it is a constraint that is always satisfied by any integral solution to the IP formulation, and that, on one hand, it can be syntactically derived by the demands' requirements in the integral lattice, but which, on the other hand, might be independent of the demands' requirements in the realm of LPs (when variables assume non-integral values). Indeed, constraint (16) might not be redundant for the relaxation $F_{LP}(\mathcal{B})$ (i.e. it might be a cutting-plane). Even more, the integrality of $F_{LP}(\mathcal{B})$ could be less than 1, while the addition of constraint (16) could result in a Linear Program with integrality gap 1.

Example 3. Continuing from Example 2, consider graph G_0. It was already observed that any solution to the covering-type problem chooses at least 2 services among $\{1, 2, 3\}$. In particular, $(\{1, 2, 3\}, 2, \mathbf{1})$ is a critical constraint for the bargaining game. Constraint $x_1 + x_2 + x_3 \geq 2$ is a redundant constraint for the IP formulation of the problem, nevertheless, the addition of the constraint to the LP relaxation improves the integrality gap from $3/4$ to 1.

For a covering-type problem $\mathcal{B}$, fix now a family of k critical constraints $\mathcal{R} = \left\{ \left(R_i, t_i, \sigma^{(i)} \right) \right\}_{i \in [k]}$. The following is a linear program relaxation to $F_{IP}(\mathcal{B})$

$$\begin{aligned}
\min \quad & \sum_{e \in E} c_e \, x_e & (F_{LP}(\mathcal{B}, \mathcal{R})) \\
\text{s.t.} \quad & \sum_{e \in T_i} \rho_{i,e} x_e \geq d_i, & \forall i \in V \\
& \sum_{e \in R_j} \sigma_e^{(j)} x_e \geq t_j, & \forall j = 1, \dots, k \\
& \mathbf{0 \leq x \leq 1}
\end{aligned}$$

As discussed earlier, the already established characterization of stable outcomes is tailored to the syntactic formulation of the covering-type problem as Integer Program $F_{IP}(\mathcal{B})$. Specific to the formulation is that feasible collections of services are determined by a number of constraints associated exactly with each of the agents. It seems a fortunate coincidence that a natural property of IP formulations of combinatorial optimization problems matches an intuitive notion of fair (stable) outcomes in bargaining games. At the same time, the bargaining game of Example 3 admits no stable solution, even though its structure is simple, and none of the agents seems to be in an advantageous position in the network. It is reasonable to assume that agents would prefer to compromise so as to be able to arrive at a (new-type of) stable solution, rather than not agreeing at all. Put it differently, $F_{LP}(\mathcal{B}, \mathcal{R})$ might have integrality gap 1, when $F_{LP}(\mathcal{B})$ does not. At the same time our findings (as well as all previously known results) pertaining to the existence of stable solutions do not apply to formulation $F_{LP}(\mathcal{B}, \mathcal{R})$, since it involves constraints that are not associated with agents. In the next section we investigate new relaxed notions of stability that are captured exactly by the integrality of $F_{LP}(\mathcal{B}, \mathcal{R})$.

On the practical side, these new notions of stability admit an interpretation according to which agents would prefer to compromise as otherwise no "fair/stable" bargaining outcome would be agreed among the players. Critical constraints describe exactly necessary conditions of feasible solutions to the combinatorial optimization problem. When such critical constraints are identified by the players, they may choose to relax their bargaining power (based on their contribution in the critical constraints) for the sake of arriving at a fair/stable bargaining outcome. This notion of fair/stable outcome is explored in the next section, and is it derived by polyhedral mechanics (same way our extended notion of stability was derived in the previous sections).

Lastly, an orthogonal question to consider, which is outside the scope of this paper, is how these critical constraints are identified. Since these constraints are exactly valid constraints for the integral hull of linear programs, an answer is given by numerous techniques proposed by well-known combinatorial optimization methods, including generic cutting planes methods, e.g. Gomory-Chvatal cuts [10, 17] (see also [26]).

4.2   Stability Based on Critical Constraints

Our main contribution in this section is the introduction of a relaxed, still natural and intuitive, notion of stability that can be characterized by the integrality of linear programs. Notably, we still study outcomes $\mathcal{F} = (A \subseteq E, \{P_{i,e}\}_{i \in V, e \in E})$ of a bargaining game $\mathcal{B} = (\mathcal{H} = (V, E), d, c, \rho)$. More specifically, we will again propose a refinement of feasible outcomes (as per Definition 1), i.e. the notion of feasibility stays invariant. The new refinement, i.e. new notion of stability, is associated with a family of k critical constraints $\mathcal{R} = \left\{ \left(R_i, t_i, \sigma^{(i)} \right) \right\}_{i \in [k]}$ for $\mathcal{B}$, which will be provably a relaxed notion of the stability of Definition 2. In what follows, and for the j-th critical constraint, we denote by $V(R_j)$ the set of agents that are served by some service in R_j, i.e. $V(R_j) = \cup_{e \in R_j} e$.

Definition 4 (R-stable Outcome). *Given a bargaining game* $\mathcal{B} = (\mathcal{H} = (V, E), d, c, \rho)$ *consider family of critical constraints* $\mathcal{R} = \left\{ \left(R_i, t_i, \sigma^{(i)} \right) \right\}_{i \in [k]}.$

A feasible outcome $\mathcal{F} = (A \subseteq E, \{P_{i,e}\}_{i \in V, e \in E})$ *is called* $\mathcal{R}$*-stable if there exist non-negative* $P_{i,e}^{ind}, P_{i,e}^{R_j}$ *such that*

$$P_{i,e} = P_{i,e}^{ind} + \sum_{j \in [k]} P_{i,e}^{R_j}$$

satisfying:

- *(Individual Greed)* $P_{i,e}^{ind} > 0$, *implies that agent* i *is tight,* $i \in e$, *and service* e *is chosen.*
- *(Collective Greed)* $P_{i,e}^{R_j} > 0$, *implies that demand* t_j *of the* j*-th critical constraint is satisfied tightly,* $e \in R_j$ *and* $i \in V(R_j)$.
- *(Collective Envy-Free) For every* $f \notin A$, *suppose that* f *contains* $\{i_1, \ldots, i_l\}$ *of the agents, as well as critical constraints* $\{j_1, \ldots, j_m\}$ *are serviced by* f, *i.e.* $f \in R_{j_t}, t = 1, \ldots, m$. *Then, for all* $e_j \in A \cap T_{i_j}$ *and for all* $e_t \in A \cap R_{j_t}$

$$\sum_{j=1}^{l} \frac{\rho_{i_j,f}}{\rho_{i_j,e_j}} P_{i_j,e_j}^{ind} + \sum_{t=1}^{m} \frac{\sigma_f^{(j_t)}}{\sigma_{e_t}^{(j_t)}} \sum_{s \in V(R_{j_t})} P_{s,e_t}^{R_{j_t}} \leq c_f. \tag{17}$$

$\mathcal{R}$-stable outcomes admit an intuitive and natural interpretation. Indeed, consider outcome $(A \subseteq E, \{P_{i,e}\}_{i \in V, e \in E})$. Each agent, as per the feasibility requirement, is still expected to make some contribution for each service $e \in A$, and the cost of the chosen services is covered by the agents. Now each agent has a relaxed notion of stability which is associated with the critical constraints. In particular, the payment of each agent i for $e \in A$ has two components, the *individual* contribution and the *collective* contribution.

More specifically, critical constraints $\mathcal{R}$ identify coalitions $V(R_j)$, $j = 1, \ldots, m$ of agents. Members within each coalition $V(r_j)$ are guaranteed to meet demand t_j with respect to satisfaction function $\sigma^{(j)}$, for every collection of services satisfying agents' demands (due to the definition of critical constraints). As such, *socially-aware* agents in $V(R_j)$ are asked to make some *collective* contribution $P_{i,e}^{R_j}$ for being members of the coalition.

On top of that, each agent makes another *individual* contribution $P_{i,e}^{ind}$ which can be thought as the only contribution agent would make in a stable outcome. Still, a notion of *fairness* is required for both types of payments to be appealing to the agents. As before, no agent should have a positive individual payment for a service e that is either not chosen in the solution or if agent is not served by e. Similarly, each agent i will have a positive collective contribution toward coalition R_j only if $i \in V(R_j)$ (i.e. only if i lies within a service required by R_j) and only when coalition R_j is satisfied tightly with respect to demand t_j and satisfaction $\sigma^{(j)}$.

Lastly, agents V and coalitions $V(R_j)$ can be thought as individual players. In particular, coalition $V(R_j)$ can be thought to make payment $\sum_{s \in V(R_j)} P_{s,e}^{R_j}$ for each $e \in E$. A natural notion of fairness then requires that each of these players makes no payment which is more than her outside option in a naturally defined "augmented" bargaining game. At a high level, each of the agents or coalitions could deviate from the current proposed selection of services, if that would result to a smaller payment, either for the agents or the coalitions. The fact that coalition-payments are distributed, in our definition of $\mathcal{R}$-stability, arbitrarily among a carefully chosen collection of agents is a natural consequence of the definition of critical constraints. This is explored in detail in the next section.

4.3 $\mathcal{R}$-Stability from $\emptyset$-Stability

In this section we justify how $\mathcal{R}$-stability is derived naturally from the notion of stability of Definition 2. First, it is not difficult to see that $\emptyset$-stable outcomes as per Definition 4 are exactly the stable outcomes of Definition 2. Next we argue that $\mathcal{R}$-stability can be derived naturally from $\emptyset$-stability for a carefully defined bargaining game.

Indeed, consider a bargaining game $\mathcal{B} = (\mathcal{H}, d, c, \rho)$, where $\mathcal{H} = (V, E), V = \{1, 2, \ldots, n\}$, along with a family of k critical constraints $\mathcal{R} = \left\{ \left(R_i, t_i, \sigma^{(i)} \right) \right\}_{i \in [k]}$. We introduce the $\mathcal{R}$-augmented bargaining game $\mathcal{B}' = (\mathcal{H}' = (V', E'), d', c', \rho')$. At a high level $\mathcal{B}'$ contains one auxiliary new agent $n+j$ for every critical constraint $R_j, j = 1, \ldots, k$. The set of services is updated so that each service contains also all of the critical constraint-players it serves. The new services have the same cost as the services in $\mathcal{B}$. Finally, the demands of the new players and their satisfactions are given by t_i and $\sigma^{(i)}$, respectively.

Formally, the $\mathcal{R}$-augmented bargaining game $\mathcal{B}'$ is defined as: $V' := V \cup \{n+1, \ldots, n+k\}$; E' contains all $e' \subseteq V'$ obtained from $e \in E$ as $e' := e \cup \{n+j : e \in R_j \text{ for some } j \in [k]\}$; $d'_i = d_i$ whenever $i \in V$, and $d'_{n+j} = t_j$ for all $j = 1, \ldots, k$; $c'_{e'} = c_e$ for all $e \in E$ (recall the two games have the same number of services); $\rho'_{i,e'} = \rho_{i,e}$ whenever $i \in V$, and $\rho'_{n+j,e'} = \sigma^{(j)}_e$ for all $j = 1, \ldots, k$. Note that by contruction, there is a bijection between E and E', and so for each $e \in E$ we denote below by $e' \in E'$ the corresponding edge in the $\mathcal{R}$-augmented bargaining game $\mathcal{B}'$.

Since $F_{LP}(\mathcal{B}, \mathcal{R})$ is a relaxation to the exact formulation $F_{IP}(\mathcal{B})$, the following is immediate from Theorem 1.

Corollary 1. $\mathcal{R}$-augmented bargaining game $\mathcal{B}'$ admits an $\emptyset$-stable (stable) outcome if and only if the integrality gap of $F_{LP}(\mathcal{B}, \mathcal{R})$ is 1.

Notably, we were able to obtain Corollary 1 only because Theorem 1 accounted for bargaining games over networks induced by hypergraphs (the augmented bargaining game involves services that are hyperedges even if the underlying graph of the original bargaining game is simple). We use the same property to obtain the main contribution of this section which reads as follows.

Theorem 3. Let $\mathcal{R}$ be a family of critical constraints for the bargaining game $\mathcal{B}$, and let $\mathcal{B}'$ be the associated $\mathcal{R}$-augmented bargaining game. Then, $\mathcal{B}$ admits an $\mathcal{R}$-stable (and feasible) solution iff $\mathcal{B}'$ admits an $\emptyset$-stable (and feasible) solution.

We prove Theorem 3 in Lemmata 8, 9 below.

Lemma 8. *If $\mathcal{B}$ admits an $\mathcal{R}$-stable (and feasible) solution then $\mathcal{B}'$ admits an $\emptyset$-stable (and feasible) solution.*

Proof. Consider an $\mathcal{R}$-stable (and feasible) solution $\mathcal{F} = (A \subseteq E, \{P_{i,e}\}_{i \in V, e \in E})$ for $\mathcal{B}$. By Definition 4, there exist non-negative $P_{i,e}^{ind}, P_{i,e}^{R_j}$ such that $P_{i,e} = P_{i,e}^{ind} + \sum_{j \in [k]} P_{i,e}^{R_j}$.

Now we define the outcome for $\mathcal{B}'$. First recall that there is a bijection between E and E', and in particular services e' in E are obtained from services in E containing (as edges) all critical constraints they serve. We set $A' = A$. For each $i \in V$ (set of original players), we set $P'_{i,e'} = P_{i,e}^{ind}$. For each critical constraint $j \in [k]$, we set $P'_{n+j,e'} = \sum_{i \in V(R_j)} P_{i,e}^{R_j}$. Next we prove that $\mathcal{F}' = (A' \subseteq E', \{P'_{i,e'}\}_{i \in V', e' \in E'})$ is feasible and $\emptyset$-stable for the $\mathcal{R}$-augmented game $\mathcal{B}'$.

Indeed, A satisfies Demand Satisfaction for $\mathcal{B}$. At the same time, we know that any collection of services with this property also satisfies any critical constraint (see Definition 3). Hence, A' satisfies Demand Satisfaction for $\mathcal{B}'$.

Now consider $e' \in A'$ (and hence $e \in A$). We verify Cost Recovery for $\mathcal{B}'$. Indeed,

$$\sum_{i \in V'} P'_{i,e'} = \sum_{i \in V} P'_{i,e'} + \sum_{j \in [k]} P'_{n+j,e'} = \sum_{i \in V} P_{i,e}^{ind} + \sum_{j \in [k]} \sum_{i \in V(R_j)} P_{i,e}^{R_j}$$

By the definition of $\mathcal{R}$-stability, $P_{i,e}^{R_j} = 0$, whenever $i \notin V(R_j)$. Hence, expression above equals

$$\sum_{i \in V} \left(P_{i,e}^{ind} + \sum_{j \in [k]} P_{i,e}^{R_j} \right) = \sum_{i \in V} P_{i,e} = c_e$$

as wanted (where the last equality is due to Cost recovery of $\mathcal{B}$). This concludes that outcome $\mathcal{F}'$ is feasible for $\mathcal{B}'$.

Now we show that $\mathcal{F}'$ is $\emptyset$-stable for $\mathcal{B}'$. First we verify property Greed. Consider $i \in V'$ and $e' \in E'$ such that $P'_{i,e'} > 0$. We consider two cases. If $i \in V$, then $P_{i,e}^{ind} > 0$. But then, Individual Greed for $\mathcal{F}$ implies that agent i is tight and $i \in e$ as wanted. If $i \in V' \setminus V$, then $i = n + j$ for some $j \in [k]$ and $P'_{n+j,e'} = \sum_{t \in V(R_j)} P_{t,e}^{R_j} > 0$. Since all payments are non-negative, there exists $t \in V(R_j)$ such that $P_{t,e}^{R_j} > 0$. Due to Collective Greed, critical constraint must be satisfied tightly, and $e' \in R_j$ (in other words, e' serves player j), as wanted.

Finally, we verify Envy-Free. Consider some $f' \notin A'$. Suppose that f' contains original agents $U \subseteq V$ and critical constraints $W \subseteq [k]$. For each $i \in U$, consider $e'_i \in A' \cap T_{i_s}$, and for each $j \in W$ consider $e'_j \in A' \cap R_j$. Taking into consideration the definition of $\mathcal{B}'$ (and the satisfaction rates of players corresponding to critical constraints), we have

$$\sum_{i \in f} \frac{\rho_{i,f'}}{\rho_{i,e'_i}} P'_{i,e'_i} = \sum_{i \in U} \frac{\rho_{i,f'}}{\rho_{i,e'_i}} P'_{i,e'_i} + \sum_{j \in W} \frac{\sigma^{(j)}_{f'}}{\sigma^{(j)}_{e'_j}} P'_{n+j,e'_j}$$

$$= \sum_{i \in U} \frac{\rho_{i,f}}{\rho_{i,e_i}} P^{ind}_{i,e_i} + \sum_{j \in W} \frac{\sigma^{(j)}_f}{\sigma^{(j)}_{e_j}} \sum_{k \in V(R_j)} P^{R_j}_{k,e_j}$$

which is at most c_f (due to that $\mathcal{F}$ satisfies the Collective Envy-Free property), as wanted.   □

Lemma 9. *If $\mathcal{B}'$ admits a stable (and feasible) solution then $\mathcal{B}$ admits an $\mathcal{R}$-stable (and feasible) solution.*

Proof. Consider outcome $\mathcal{F}' = \left(A' \subseteq E', \{P'_{i,e'}\}_{i \in V', e' \in E'}\right)$ for bargaining game $\mathcal{B}'$, which is feasible and stable.

For each critical constraint $j \in [k]$, we consider an arbitrary distribution $\{\tau^{(j)}_i\}_{i \in V(R_j)}$ over agents in $V(R_j)$, i.e. $\sum_{i \in V(R_j)} \tau^{(j)}_i = 1$, with all $\tau^{(j)}_i \geq 0$. For completeness, we also set $\tau^{(j)}_i = 0$ whenever $i \notin R_j$. Next we define outcome $\mathcal{F} = (A \subseteq E, \{P_{i,e}\}_{i \in V, e \in E})$ for bargaining game $\mathcal{B}$ as follows; $A = A'$, and for every $i \in V$ and $e \in E$, we set $P^{ind}_{i,e} = P'_{i,e'}$ and for each critical constraint $j \in [k]$ we set $P^{R_j}_{i,e} = \tau^{(j)}_i P'_{n+j,e'}$, so that

$$P_{i,e} = P'_{i,e'} + \sum_{j \in [k]} \tau^{(j)}_i P'_{n+j,e'}.$$

Next we show that $\mathcal{F}$ is feasible and $\mathcal{R}$-stable.

First note that A' by definition satisfies all agents' demands in $\mathcal{B}'$ (even the demands of the critical constraints), hence A satisfies demand satisfaction for $\mathcal{B}$. Next we study Cost Recovery, so we consider some arbitrary $e \in A$ (hence $e' \in A'$). We have

$$\sum_{i \in V} P_{i,e} = \sum_{i \in V} \left(P'_{i,e'} + \sum_{j \in [k]} \tau^{(j)}_i P'_{n+j,e'} \right)$$

$$= \sum_{i \in V} P'_{i,e'} + \sum_{j \in [k]} P'_{n+j,e'} \sum_{i \in V} \tau^{(j)}_i$$

$$= \sum_{i \in V} P'_{i,e'} + \sum_{j \in [k]} P'_{n+j,e'}$$

$$= c_e$$

where the last equality is due to that $\mathcal{F}'$ satisfies Cost recovery for $\mathcal{B}'$.

In what follows we show that $\mathcal{F}$ is $\mathcal{R}$-stable. First we study Individual Greed. Consider $i \in V, e \in E$ so that $P^{ind}_{i,e} > 0$. But then, $P'_{i,e'} > 0$ in $\mathcal{B}'$, and since $\mathcal{F}'$ is stable, we conclude that $i \in e$ and i is satisfied tightly. as wanted.

Next we study Collective Greed. Consider $i \in V, e \in E, j \in [k]$ so that $P^{R_j}_{i,e} > 0$. But then, $\tau^{(j)}_i P'_{n+j,e'} > 0$ in $\mathcal{B}'$. Recall that distribution $\{\tau^{(j)}_i\}_i$ has its support within $V(R_j)$, hence $i \in V(R_j)$. Moreover, $P'_{n+j,e'} > 0$ implies that $e' \in A' \cap R_j$ and critical constraint R_j has its demand satisfied tightly, since $\mathcal{F}'$ is stable. Hence $\mathcal{F}$ satisfies Collective Greed.

Finally, we show that $\mathcal{F}$ is Collective Envy-Free. Indeed, consider $f \notin A$ (hence $f' \notin A'$). Suppose that f contains $\{i_1, \ldots, i_l\}$ of the agents, as well as critical constraints $\{j_1, \ldots, j_m\}$ are serviced by f. Consider also arbitrary $e_j \in A \cap T_{i_j}$ and $e_t \in A \cap R_{j_t}$ (hence $e'_j, e'_t \in A'$). Then,

$$\sum_{j=1}^{l} \frac{\rho_{i_j,f}}{\rho_{i_j,e_j}} P^{ind}_{i_j,e_j} + \sum_{t=1}^{m} \frac{\sigma_f^{(j_t)}}{\sigma_{e_t}^{(j_t)}} \sum_{s \in V(R_{j_t})} P^{R_{j_t}}_{s,e_t}$$

$$= \sum_{j=1}^{l} \frac{\rho_{i_j,f}}{\rho_{i_j,e_j}} P'_{i_j,e'_j} + \sum_{t=1}^{m} \frac{\sigma_f^{(j_t)}}{\sigma_{e_t}^{(j_t)}} \sum_{s \in V(R_{j_t})} \tau_s^{(j_t)} P'_{n+j_t,e'_t}$$

$$= \sum_{j=1}^{l} \frac{\rho_{i_j,f}}{\rho_{i_j,e_j}} P'_{i_j,e'_j} + \sum_{t=1}^{m} \frac{\sigma_f^{(j_t)}}{\sigma_{e_t}^{(j_t)}} P'_{n+j_t,e'_t} \sum_{s \in V(R_{j_t})} \tau_s^{(j_t)}$$

Recalling that $\{\tau_s^{(j_t)}\}_s$ is a distribution over $s \in V(R_{j_t})$, the last expression is less than c_f, since $\mathcal{F}'$ is Envy-Free for $\mathcal{B}'$.   □

Corollary 1 together with Theorem 3 characterize the existence of $\mathcal{R}$-stable outcomes.

Corollary 2. *Bargaining game $\mathcal{B}$ admits an $\mathcal{R}$-stable outcome if and only if the integrality gap of $F_{LP}(\mathcal{B}, \mathcal{R})$ is 1.*

We conclude this section by demonstrating an application of our findings pertaining to $\mathcal{R}$-stable outcomes.

Example 4. Consider the covering-type problem of Example 1, and fix graph $G = (V_0, E_0)$. A number of cutting planes are known for the problem, including the so-called odd-cycle constraints; for every odd cycle $C \subseteq V$, and for every vertex cover A of G, we have $|A \cap C| \geq (|C| + 1)/2$. Effectively, cutting planes $\sum_{i \in C} x_i \geq (|C|+1)/2$, for all odd cycle $C \subseteq V$ can be added to formulation (2). Moreover each odd cycle C is a critical constraint. The specific graph of bargaining game of Example 2 did not admit a stable solution. However, together with the odd-cycle (critical) constraint $R = (\{1,2,3\}, 2, 1)$, and as explained in Example 3, the new LP relaxation has integrality gap 1. Hence it admits an $\mathcal{R}$-stable solution, in which each *socially-aware* agent not only makes an individual payment, but also contributes towards covering the cost of the global solution as a member of a coalition (the odd cycle) which always receives at least 2 services.

5   Discussion and Open Problems

We studied covering-type problems and their underlying bargaining games over networks. Our work generalized previously known results in two directions. First, our networks are induced by hypergraphs, and hence bargaining is associated with generic subsets of players (not only of size 2). Second, we assumed non-uniformity for the "value" of the objects to be bargained over, in contrast to previous results in which all objects were worth the same to all agents. These generalizations allowed us to further extend previous results by introducing and characterizing new and relaxed notions of stability based on combinatorial properties of the underlying optimization problems, which admit intuitive interpretations based on socially-aware players.

Our work is a just starting point toward introducing natural notions of stability based on properties of the underlying combinatorial optimization problems. Indeed, we introduced intuitive notions of stability based on linear programs strengthened by constraints valid for the integral hull of the relaxations, and we only considered linear tightenings. Is there a natural stability notion associated with semidefinite programming tightenings (or in general non-linear tightenings)? What if the starting linear program relaxation is tightened by any of the well-studied lift-and-project systems, e.g. Lovasz-Schrijver [24], Sherali-Adams [29], or Lasserre [23]? Similarly, is there a natural stability notion associated with convex relaxations in which the integrality gap is not 1 but bounded?

References

1. Azar, Y., Birnbaum, B., Celis, L.E., Devanur, N.R., Peres, Y.: Convergence of local dynamics to balanced outcomes in exchange networks. In: 50th Annual IEEE Symposium on Foundations of Computer Science, FOCS 2009, Atlanta, Georgia, USA, 25–27 October 2009, pp. 293–302. IEEE Computer Society (2009)
2. Bateni, M.H., Hajiaghayi, M.T., Immorlica, N., Mahini, H.: The cooperative game theory foundations of network bargaining games. In: Abramsky, S., Gavoille, C., Kirchner, C., Meyer auf der Heide, F., Spirakis, P.G. (eds.) ICALP 2010. LNCS, vol. 6198, pp. 67–78. Springer, Heidelberg (2010). https://doi.org/10.1007/978-3-642-14165-2_7
3. Bayati, M., Borgs, C., Chayes, J., Kanoria, Y., Montanari, A.: Bargaining dynamics in exchange networks. J. Econ. Theory **156**, 417–454 (2015)
4. Bock, A., Chandrasekaran, K., Könemann, J., Peis, B., Sanità, L.: Finding small stabilizers for unstable graphs. Math. Program. **154**(1–2), 173–196 (2015)
5. Celis, L.E., Devanur, N.R., Peres, Y.: Local dynamics in bargaining networks via random-turn games. In: Saberi, A. (ed.) WINE 2010. LNCS, vol. 6484, pp. 133–144. Springer, Heidelberg (2010). https://doi.org/10.1007/978-3-642-17572-5_11
6. Chakraborty, T., Kearns, M.: Bargaining solutions in a social network. In: Papadimitriou, C., Zhang, S. (eds.) WINE 2008. LNCS, vol. 5385, pp. 548–555. Springer, Heidelberg (2008). https://doi.org/10.1007/978-3-540-92185-1_61
7. Chakraborty, T., Kearns, M., Khanna, S.: Network bargaining: algorithms and structural results. In: Proceedings 10th ACM Conference on Electronic Commerce, EC 2009, Stanford, California, USA, 6–10 July 2009, pp. 159–168 (2009)
8. Chalkiadakis, G., Elkind, E., Wooldridge, M.: Computational Aspects of Cooperative Game Theory. Morgan & Claypool Publishers, San Rafael (2011)

9. Chan, T.-H.H., Chen, F., Ning, L.: Optimizing social welfare for network bargaining games in the face of unstability, greed and spite. In: Epstein, L., Ferragina, P. (eds.) ESA 2012. LNCS, vol. 7501, pp. 265–276. Springer, Heidelberg (2012). https://doi.org/10.1007/978-3-642-33090-2_24

10. Chvátal, V.: Edmonds polytopes and a hierarchy of combinatorial problems. Discrete Math. **306**(10–11), 886–904 (2006)

11. Cook, K.S., Emerson, R.M., Gillmore, M.R., Yamagishi, T.: Distribution of power in exchange networks: theory and experimental results. Am. J. Sociol. **89**, 275–305 (1983)

12. Deng, X., Fang, Q.: Algorithmic cooperative game theory. In: Chinchuluun, A., Pardalos, P.M., Migdalas, A., Pitsoulis, L. (eds.) Pareto Optimality, Game Theory and Equilibria. SOIA, vol. 17, pp. 159–185. Springer, New York (2008). https://doi.org/10.1007/978-0-387-77247-9_7

13. Deng, X., Ibaraki, T., Nagamochi, H.: Algorithmic aspects of the core of combinatorial optimization games. Math. Oper. Res. **24**(3), 751–766 (1999)

14. Draief, M., Vojnovic, M.: Bargaining dynamics in exchange networks. CoRR, abs/1202.1089 (2012)

15. Farczadi, L., Georgiou, K., Könemann, J.: Network bargaining with general capacities. In: Bodlaender, H.L., Italiano, G.F. (eds.) ESA 2013. LNCS, vol. 8125, pp. 433–444. Springer, Heidelberg (2013). https://doi.org/10.1007/978-3-642-40450-4_37

16. Georgiou, K., Karakostas, G., Könemann, J., Stamirowska, Z.: Social exchange networks with distant bargaining. Theoret. Comput. Sci. **554**, 263–274 (2014)

17. Gomory, R.E.: Solving linear programming problems in integers. In: Bellman, R., Hall Jr., M. (eds.) Combinatorial Analysis, pp. 211–215, Providence, RI. Symposia in Applied Mathematics X. American Mathematical Society (1960)

18. Granot, D., Granot, F.: On some network flow games. Math. Oper. Res. **17**(4), 792–841 (1992)

19. Hajiaghayi, M., Mahini, H.: Bargaining networks. In: Kao, M.Y. (ed.) Encyclopedia of Algorithms, pp. 1–5. Springer, Boston (2014). https://doi.org/10.1007/978-3-642-27848-8

20. Ito, T., Kakimura, N., Kamiyama, N., Kobayashi, Y., Okamoto, Y.: Efficient stabilization of cooperative matching games. Theoret. Comput. Sci. **677**, 69–82 (2017)

21. Kanoria, Y., Bayati, M., Borgs, C., Chayes, J.T., Montanari, A.: A natural dynamics for bargaining on exchange networks. CoRR, abs/0911.1767 (2009)

22. Kleinberg, J.M., Tardos, É.: Balanced outcomes in social exchange networks. In: Proceedings of the ACM Symposium on Theory of Computing, pp. 295–304 (2008)

23. Lasserre, J.B.: An explicit exact SDP relaxation for nonlinear 0-1 programs. In: Aardal, K., Gerards, B. (eds.) IPCO 2001. LNCS, vol. 2081, pp. 293–303. Springer, Heidelberg (2001). https://doi.org/10.1007/3-540-45535-3_23

24. Lovász, L., Schrijver, A.: Cones of matrices and set-functions and 0-1 optimization. SIAM J. Optim. **1**(2), 166–190 (1991)

25. Nash, J.: The bargaining problem. Econometrica **18**, 155–162 (1950)

26. Nemhauser, G.L., Wolsey, L.A.: Integer and Combinatorial Optimization. Wiley, Hoboken (1988)

27. Rochford, S.C.: Symmetrically pairwise-bargained allocations in an assignment market. J. Econ. Theory **34**(2), 262–281 (1984)

28. Shapley, L.S., Shubik, M.: The assignment game: the core. Int. J. Game Theory **1**(1), 111–130 (1971)
29. Sherali, H.D., Adams, W.P.: A hierarchy of relaxations between the continuous and convex hull representations for zero-one programming problems. SIAM J. Discrete Math. **3**(3), 411–430 (1990)
30. Willer, D.: Network Exchange Theory. Praeger Publishers, Westport (1999)

Consensus Reaching with Heterogeneous User Preferences

Hélène Le Cadre[(✉)], Enrique Rivero Puente, and Hanspeter Höschle

VITO/EnergyVille, Thor Scientific Park, 3600 Genk, Belgium
`helene.lecadre@energyville.be`

Abstract. In this paper, we consider consumers and prosumers who interact on a platform. Consumers buy energy to the platform to maximize their usage benefit while minimizing the cost paid to the platform. Prosumers, who have the possibility to generate energy, self-consume part of it to maximize their usage benefit and sell the rest to the platform to maximize their revenue. Product differentiation is introduced and consumers can specify preferences regarding locality, RES-based generation, and matchings with the prosumers. The consumers and prosumers' problems being coupled through a matching probability, we provide analytical characterizations of the resulting Nash equilibrium. Assuming supply-shortages occur, we reformulate the platform problem as a consensus problem that we solve using Alternating Direction Method of Multipliers (ADMM), enabling minimal information exchanges between the nodes. On top of the platform, a trust-based mechanism combining exploitation of nodes with good reputation and exploration of new nodes, is implemented to determine the miner node which validates the transactions. A case study is provided to analyze the impact of preferences and miner selection dynamic process.

Keywords: Game theory · Two-sided market · Local community · ADMM

1 Introduction

The increasing amount of Distributed Energy Resources (DERs), which have recently been integrated in power systems, the development of new storage technologies, and the more proactive role of consumers (prosumers) have transformed the classical centralized power system operation (mostly based on unit commitment) by introducing more uncertainty and decentralization in the decisions. Following this trend, electricity markets are starting to restructure, from a centralized market design in which all the operations were managed by a global (central) market operator, modeled as a classical constrained optimization problem, to more decentralized designs involving local energy communities which can trade energy by the intermediate of the global market operator [6] or, in a peer-to-peer setting [8]. Coordinating local renewable energy sources (RES)-based

© ICST Institute for Computer Sciences, Social Informatics and Telecommunications Engineering 2019
Published by Springer Nature Switzerland AG 2019. All Rights Reserved
K. Avrachenkov et al. (Eds.): GameNets 2019, LNICST 277, pp. 151–170, 2019.
https://doi.org/10.1007/978-3-030-16989-3_11

generators to satisfy the demand of local energy communities, could provide significant value to the power systems, by decreasing the need for investments in conventional generations and transmission networks. In practice, the radial structure of the distribution grid calls for hierarchical market designs, involving transmission and distribution network operators [7]. But, various degrees of coordination can be envisaged, from full coordination organized by a global market operator (transmission network operator), to bilateral contract networks, to fully decentralized market designs allowing peer-to-peer energy trading between the prosumers in a distributed fashion [8], or within and between coalitions of prosumers [12].

In the energy sector, peer-to-peer energy trading is a novel paradigm of power system operation [14], where prosumers providing their own energy from solar panels, storage technologies, demand response mechanisms, exchange energy/capacity with one another. Zhang et al. provide in [15] an exhaustive list of projects and trails all around the world, which build on new innovative approaches for peer-to-peer energy trading. A large part of these projects rely on *platforms* which match RES-based generators and consumers according to their preferences and locality aspects (such as Piclo in the UK, TransActive Grid in Brooklyn, US, Vandebron in the Netherlands, etc.). In the same vein, cloud-based virtual market places to deal with excess generation within microgrids are developed by PeerEnergyCloud and Smart Watts in Germany. Some other projects rely on local community building for investment sharing in batteries, solar PV panels, etc., in exchange of bill reduction or to obtain a certain level of autonomy with respect to the global grid (such as Yeloha and Mosaic in the US, SonnenCommunity in Germany, etc.).

Platform design is an active area of research in the two-sided market economics literature [3]. Three needs are identified for platform deployment: a first requirement is to help buyers and sellers find each other, taking into account preference heterogeneity. This requires to find a trade-off between low-entry cost and information retrieval from big, heterogeneous, and dynamic information flows. Buyers and sellers search can be performed in a centralized fashion (Amazon, Uber), or it might allow for effective decentralized search (Airbnb, eBay), or even fully distributed search (OpenBazaar, Arcade City). A second need is to set prices that balance demand and supply, and ensure that prices are set competitively in a decentralized fashion. A third requirement is to maintain trust in the market, relying on reputation and feedback mechanisms. Sometimes, supply might be insufficient and subsidies should be designed to encourage sharing on the platform [3].

From an information and communication technology (ICT) perspective, a fully decentralized market design provides a robust framework since if one node in a local market is attacked or in case of failures, the whole architecture should remain in place and information could find other paths to circulate from one point to another, avoiding malicious nodes/corrupted paths. From an algorithmic point of view, such a setting enables the implementation of algorithms that preserve privacy of the local market agents (requiring from them to not share more than their dual variables - e.g., local prices - updates). This also creates high computational challenges, especially if the number of local markets/peers is high. Trust, security, and transparency issues for peer-to-peer energy markets could rely on the emergence of blockchain technology. A blockchain is a continuously growing list of records, called blocks, which are linked and secured using cryptography. Each block typically contains a hash pointer as a link to a previous block, a time-stamp and transaction data. By design, blockchains are inherently resistant to modification of the data. A blockchain can serve as "an open, distributed ledger that can record transactions between two or more parties efficiently and in a verifiable and permanent way". For use as a distributed ledger, a blockchain is typically managed by a peer-to-peer network collectively adhering to a protocol for validating new blocks. Once recorded, the data in any given block cannot be altered retroactively without the alteration of all subsequent blocks, which needs a collusion of the network majority. The most important function about the records (called transactions in the literature) is their traceability. For each record, it is possible to trace its origin and by whom it has been created and/or exchanged. The verification of the correctness of each transaction could be done by every participant of the chain. However, there is a specific role for creating a block at every time period and thus, guaranteeing that the transactions within it are correct. This is the role of the so-called *miner*, who provides computational power to check the transactions and put them together to form blocks, in exchange for a fee [13]. On top of blockchain technology, smart contracts are autonomous computer systems, written in code, that manage executions in the form of rules between parties on the Blockchain. For example, the reaching of a consensus between nodes, specific events (train and airplane delays, conditions for a contract to hold) can be detected online, and the execution of the smart contract is automatically triggered [11].

To avoid any influence of a malicious node, consensus algorithms are employed. Bitcoin, the first existing blockchain technology, relies on Proof-of-Work (PoW): nodes have to solve a mathematical problem (puzzle) so complex that the only way to solve it is to try every possible permutation. This results in a slow and excessively energy-greedy program. Proof-of-Stake (PoS) used by Etherum, another well-known blockchain technology, is a method for consensus-building between the nodes, in which a miner is randomly selected based on its wealth instead of reputation. However, the miner node selection seems rather arbitray and not relying on trust. In [9], Munsing et al. consider a large-scale load scheduling problem that they decompose with ADMM [1, 4]. Consensus is reached when ADMM converges to a stable solution; payments and penalties are then computed based on the ADMM outcome. ADMM-based consensus problems consist in computing the optimum of a (large-scale) optimization problem, where nodes exchange information only with a subset of the other nodes. Several algorithms for consensus can be found in the literature and have attracted much attention in the last decades in the broader framework of sensor management and data fusion: they differentiate on the basis of the amount of communication and computation they use, on their scalability with respect to the number of nodes, on their (online) adaptability, and, finally, they can be deterministic or randomized [2].

The contributions of this manuscript can be summarized as follows:

(a) We formulate an equilibrium problem representing a two-sided local market platform with consumers and prosumers. Consumers can have preferences toward characteristics of the electricity products (RES-based generation, locality) and matchings with the prosumers.

(b) We discuss different algorithms to compute the outcome of such a market platform and highlight differences in the need of sharing information with the local market operator and other participants.

(c) We propose a method on how the role of the market operator could be organized with smart contracts and the miner node can be selected among the market participants relying on a trust-based mechanism.

(d) We illustrate (b) and (c) based on a small case study highlighting the effect of certain parameter, most importantly, the consumer preferences, and the tuning of the parameters for the trust-based mechanism.

The remainder of the paper is structured as follows. Section 2 introduces the mathematical description of the market platform and its participants. The proposed ADMM algorithm to compute the clearing of the market platform is introduced in Sect. 3. Section 4 illustrates and discusses several elements of the market platform in a case study. Conclusions of the paper are drawn in Sect. 5.

2   Model Description

We consider a set $\mathcal{N}$ of N nodes. Each node can be either a *prosumer* P having the possibility to generate and consume (part) of her own energy while selling the excess by the intermediate of a sharing platform operated by a local Market Operator (MO), or a *consumer-only* C without generation facility. We denote by $\mathcal{P}$, the prosumer set, and by $\mathcal{C}$, the consumer-only set. Furthermore, we have the relations: $\mathcal{P} \cup \mathcal{C} = \mathcal{N}$ and $\mathcal{C} \cap \mathcal{N} = \emptyset$. Local energy demand and supply balance is guaranteed by the local MO, who can sell excess production or buy shortage to the power grid.

Our inspiration for the prosumer-consumer interaction model comes for the literature of two-sided markets [3], though the structure of electricity markets and asymmetry of prosumer role, who can benefit from consumption of self-production (therefore, becoming consumers) and excess production selling by the intermediate of the sharing platform (therefore, becoming producers), makes extensions of this literature tricky. The consumer-prosumer platform framework is visualized in Fig. 1.

2.1   Modeling Consumers

For each consumer $C \in \mathcal{C}$, we denote the usage benefit obtained from consuming a quantity y_t^C of energy, by $U_C(y_t^C)$. We assume that $U_C(.)$ is only known to

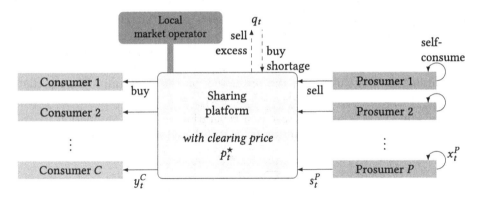

Fig. 1. Example of a sharing platform involving consumers-only on one side and prosumers on the other side.

the consumer and is not public knowledge. We make the assumption that $U_C(.)$ is continuous and strictly concave and non-negative on $\mathbb{R}_+$. To fix the idea, we assume that consumer C usage benefit is a quadratic function of the consumer demand y_t^C, leading to the following definition:

$$U_C(y_t^C) = -\eta^C(y_t^C - y_t^{C\sharp})^2 + \tilde{\eta}^C, \tag{1}$$

where $\eta^C, \tilde{\eta}^C$ are positive parameters, and $y_t^{C\sharp}$ is the target demand of consumer C at time period t. For the usage benefit to remain non-negative on the interval of definition of y_t^C, we impose conditions on the parameters such that $U_C(0) \geq 0$ and $U_C(\kappa^C) \geq 0$, leading to $\kappa^C - \sqrt{\frac{\tilde{\eta}^C}{\eta^C}} \leq y_t^{C\sharp} \leq \sqrt{\frac{\tilde{\eta}^C}{\eta^C}}, \forall t$. Note that the maximum usage benefit is reached in $U_C(y_t^{C\sharp}) = \tilde{\eta}^C$ and in case $U_C(0) = 0$, i.e., zero demand implies zero usage benefit, we have the following relation between the consumer target demand and usage benefit parameters: $\eta^C = \frac{U_C(y_t^{C\sharp})}{(y_t^{C\sharp})^2}$.

We refine the consumer model by introducing product differentiation [8]. To that purpose, we first assume that consumer C defines the percentages ξ_{RES}^C, $\xi_{\text{Loc}}^C \in [0; 1]$ of his target demand that comes from "RES-based generation" and local energy prosumers, where the "locality" of prosumer P with respect to consumer C is measured by the distance between P and C^1. So, we assume that some consumers might prefer to be served by prosumers in a local area, e.g., within a limited radius. This means that the percentage of consumer C target demand coming from RES-based generation and local prosumers is defined as $\xi_{\text{RES}}^C y_t^{C\sharp}$ and $\xi_{\text{Loc}}^C y_t^{C\sharp}$ respectively. Note that we do not impose that $\xi_{\text{RES}}^C + \xi_{\text{Loc}}^C = 1$ because "RES-based generation" and "locality" are not mutually exclusive preferences, meaning that some "green" consumers might want to cover their demand with 100% RES-based generation and local production only. We add a second level of complexity, by assuming that consumer C has intrinsic preferences with

---

[1] Note that the preference model is generic enough to introduce other levels of product differentiation.

respect to the prosumers in $\mathcal{P}$ (that we will call later on "matching preferences"), which lead them to define how they ideally wish to split their demand between the prosumers. Let $\alpha_{\mathrm{RES}}^{CP}$ and $\alpha_{\mathrm{Loc}}^{CP}$ be positive parameters associated with any prosumer $P \in \mathcal{P}$ production, characterizing the preferences of the consumer C regarding his matching with the prosumers on the platform, and such that $\frac{1}{card(\mathcal{P})\xi_{\mathrm{RES}}^{C}} \sum_{P \in \mathcal{P}} \alpha_{\mathrm{RES}}^{CP} = \frac{1}{card(\mathcal{P})\xi_{\mathrm{Loc}}^{C}} \sum_{P \in \mathcal{P}} \alpha_{\mathrm{Loc}}^{CP} = 1, \forall C \in \mathcal{C}$.

Extending Eq. (1) to the RES-based generation and local production usage, we define the usage benefit resulting from RES-based and local consumption as follows:

$$U_C^{\mathrm{Loc}}(y_t^C) = -\eta_{\mathrm{Loc}}^C \Big(\sum_{P \in \mathcal{P}} \alpha_{\mathrm{Loc}}^{CP} y_t^C - \xi_{\mathrm{Loc}}^C y_t^{C\sharp} \Big)^2 + \tilde{\eta}_{\mathrm{Loc}}^C.$$

$$U_C^{\mathrm{RES}}(y_t^C) = -\eta_{\mathrm{RES}}^C \Big(\sum_{P \in \mathcal{P}} \alpha_{\mathrm{RES}}^{CP} y_t^C - \xi_{\mathrm{RES}}^C y_t^{C\sharp} \Big)^2 + \tilde{\eta}_{\mathrm{RES}}^C.$$

We introduce w_{Loc}, w_{RES}, and w_0, as non-negative parameters characterizing the relative importance of locality and RES-based generation in the consumer total usage benefit, with respect to the no-product differentiation case. Then, consumer C total usage benefit can be decomposed as the weighted sum of the benefits retrieved from local production and RES-based generation consumption, and usage benefit (1) without product differentiation[2]:

$$\tilde{U}_C(y_t^C) = w_{\mathrm{Loc}} U_C^{\mathrm{Loc}}(y_t^C) + w_{\mathrm{RES}} U_C^{\mathrm{RES}}(y_t^C) + w_0 U_C(y_t^C). \tag{2}$$

The utility consumer C obtains from energy consumption y_t^C, $\Pi_C(y_t^C)$, is given by the benefit $\tilde{U}_C(.)$ minus the cost to buy energy on the platform operated by the local market operator (MO), $p_t^\star$ times the consumption y_t^C. Formally, we have:

$$\Pi_C(y_t^C) = \tilde{U}_C(y_t^C) - p_t^\star y_t^C. \tag{3}$$

Each consumer C determines his demand y_t^C so as to maximize the sum of his utility function (3) and potential mining fee, under non-negativity and maximum capacity of consumption κ^C constraints:

$$\max_{y_t^C} \; \Pi_C(y_t^C), \tag{4}$$

$$s.t. \; y_t^C \leq \kappa^C, \qquad (\psi_t^C) \tag{5}$$

$$0 \leq y_t^C. \qquad (\tilde{\psi}_t^C) \tag{6}$$

We prove in the proposition below that there always exists a solution to the consumer utility maximization problem.

---

[2] Note that (1) is added in Eq. (2) to counter-balance the effects of product differentiation that might encourage the consumer to excess her demand in case where the sum of the locality and RES-based generation demand targets is larger ($\xi_{\mathrm{RES}}^C + \xi_{\mathrm{Loc}}^C > 1$) than the actual demand target.

Proposition 1. *Consumer C utility function $\Pi_C(.)$ is strictly concave in his demand y_t^C and maximized at a single optimum solution of optimization problem (4) under constraints (5), (6):*

$$
y_t^{C\star} = \frac{\left(2w_0\eta^C + 2w_{RES}\eta_{RES}\xi_{RES}^C(\sum_{P\in\mathcal{P}}\alpha_{RES}^{CP})\right)y_t^{C\sharp}}{2w_0\eta^C + 2w_{RES}\eta_{RES}^C(\sum_{P\in\mathcal{P}}\alpha_{RES}^{CP})^2 + 2w_{Loc}\eta_{Loc}^C(\sum_{P\in\mathcal{P}}\alpha_{Loc}^{CP})^2}
$$
$$
+ \frac{\left(2w_{Loc}\eta_{Loc}^C\xi_{Loc}^C(\sum_{P\in\mathcal{P}}\alpha_{Loc}^{CP})\right)y_t^{C\sharp} - p_t^\star - (\tilde{\varPsi}_t^C - \varPsi_t^C)}{2w_0\eta^C + 2w_{RES}\eta_{RES}^C(\sum_{P\in\mathcal{P}}\alpha_{RES}^{CP})^2 + 2w_{Loc}\eta_{Loc}^C(\sum_{P\in\mathcal{P}}\alpha_{Loc}^{CP})^2}, \quad (7)
$$

with $\varPsi_t^C(y_t^{C\star} - \kappa^C) = 0$, $\tilde{\varPsi}_t^C y_t^{C\star} = 0$, $\varPsi_t^C \geq 0$, $\tilde{\varPsi}_t^C \geq 0$.

Proof. The Lagrangian function associated with optimization problem (4)–(6) writes down as follows: $\mathcal{L}_C(y_t^C, \varPsi_t^C, \tilde{\varPsi}_t^C) = \Pi_C(y_t^C) - \varPsi_t^C(y_t^C - \kappa^C) + \tilde{\varPsi}_t^C y_t^C$. Complementarity slackness conditions take the form: $\varPsi_t^C(y_t^C - \kappa^C) = 0$, $\tilde{\varPsi}_t^C y_t^C = 0$, and the dual feasibility constraints impose that: $\varPsi_t^C \geq 0$, $\tilde{\varPsi}_t^C \geq 0$.

Derivating the Lagrangian function with respect to y_t^C, we obtain:

$$
\frac{\partial\mathcal{L}_C(y_t^C, \varPsi_t^C, \tilde{\varPsi}_t^C)}{\partial y_t^C} = -2w_0\eta^C(y_t^C - y_t^{C\sharp}) - 2w_{RES}\eta_{RES}^C(\sum_{P\in\mathcal{P}}\alpha_{RES}^{CP}y_t^C - \xi_{RES}^C y_t^{C\sharp})
$$
$$
(\sum_{P\in\mathcal{P}}\alpha_{RES}^{CP}) - 2w_{Loc}\eta_{Loc}^C(\sum_{P\in\mathcal{P}}\alpha_{Loc}^{CP}y_t^C - \xi_{Loc}^C y_t^{C\sharp})(\sum_{P\in\mathcal{P}}\alpha_{Loc}^{CP}) - p_t^\star - \varPsi_t^C + \tilde{\varPsi}_t^C.
$$

Derivating the Lagrangian function twice with respect to y_t^C, we obtain:

$$
-2w_0\eta^C - 2w_{RES}\eta_{RES}^C(\sum_{P\in\mathcal{P}}\alpha_{RES}^{CP})^2 - 2w_{Loc}\eta_{Loc}^C(\sum_{P\in\mathcal{P}}\alpha_{Loc}^{CP})^2 < 0.
$$

This implies that $\Pi_C(.)$ is strictly concave in y_t^C. Therefore, it admits a unique optimum. At the optimum in $y_t^{C\star}$, $\frac{\partial\mathcal{L}_C(y_t^C, \varPsi^C, \tilde{\varPsi}^C)}{\partial y_t^C}\big|_{y_t^C = y_t^{C\star}} = 0$, which is equivalent to (7). $\qquad\square$

2.2   Modeling Prosumers

Prosumers have two ways to derive benefits from their production: using it themselves or selling it through the sharing platform by the intermediate of the local MO. We let x_t^P be prosumer P self-usage quantity and s_t^P be the quantity of energy that prosumer P shares through the platform. When prosumers consume their own energy production, they experience benefit from the consumption, like consumers-only. But, unlike consumers-only, they do not have to pay the local MO for their consumption, though their consumption may lead to production costs that can be interpreted as usage (in case of micro-CHP activation for example) or maintenance cost, or government taxes, etc. We denote the benefit from self-usage by $U_P(x_t^P)$ and the production cost incurred by $c_P(x_t^P + s_t^P)$. As in the case of the consumers-only, we assume that $U_P(.)$ is continuous and strictly concave and non-negative on $\mathbb{R}_+$. In the same spirit as the consumer model, we assume that prosumer P usage benefit is a quadratic function of the prosumer self-consumption x_t^P, leading to the following definition:

$$U_P(x_t^P) = -\eta^P (x_t^P - x_t^{P\sharp})^2 + \tilde{\eta}^P, \tag{8}$$

where $\eta^P, \tilde{\eta}^P$ are non-negative parameters, and $x_t^{P\sharp}$ is the target self-consumption of prosumer P at time period t. For the self-consumption benefit to remain non-negative on the interval of definition of x_t^P, we impose conditions on the parameters such that $U_P(0) \geq 0$ and $U_P(\kappa^P) \geq 0$, leading to $\kappa^P - \sqrt{\frac{\tilde{\eta}^P}{\eta^P}} \leq x_t^{P\sharp} \leq \sqrt{\frac{\tilde{\eta}^P}{\eta^P}}, \forall t$.

When the prosumers share their excess production through the platform, they receive a revenue and incur costs. The revenue they receive from sharing depends on how many other prosumers are also sharing their excess production. We introduce the probability $\mu(\mathbf{y}_t, \mathbf{s}_t)$ that a prosumer is matched to a consumer-only as follows:

$$\mu(\mathbf{y}_t, \mathbf{s}_t) := \min \left\{ \frac{\sum_{C \in \mathcal{C}} y_t^C}{\sum_{P \in \mathcal{P}} s_t^P}; 1 \right\}. \tag{9}$$

Naturally, $\mu(\mathbf{y}_t, \mathbf{s}_t) < 1$ if, and only if, $\sum_{C \in \mathcal{C}} y_t^C < \sum_{P \in \mathcal{P}} s_t^P$, i.e., there is an excess of supply compared to the actual demand on the platform. And, $\mu(\mathbf{y}_t, \mathbf{s}_t) = 1$ in case the consumer total demand is larger than the prosumers supply, therefore requiring that the local MO buys the missing quantity to the grid. In the following, for the sake of simplicity, we will write: $\mu_t := \mu(\mathbf{y}_t, \mathbf{s}_t)$.

The utility function of a prosumer is the sum of the benefit she derives from the consumption of her self-production plus the expected revenue she derives from the selling of her excess production conditionally to her matching with a consumer minus her production cost, leading to the following mathematical expression:

$$\Pi_P(x_t^P, \mathbf{y}_t, \mathbf{s}_t) = U_p(x_t^P) + p_t^\star \mu_t s_t^P - c_P(x_t^P + s_t^P). \tag{10}$$

Assuming that prosumer P cost function is quadratic in her production, we set $c_P(x) = c_{P2}x^2 + c_{P1}x + c_{P0}, \forall x \in \mathbb{R}$ with c_{P2}, c_{P1}, c_{P0} non-negative parameters.

Each prosumer P determines sharing and self-use variables s_t^P and x_t^P that maximize the sum of her utility function (10) and potential mining fee, under maximum capacity of production κ^P constraint, by solving the following optimization problem:

$$\max_{x_t^P, s_t^P} \Pi_P(x_t^P, \mathbf{y}_t, \mathbf{s}_t), \tag{11}$$

$$\text{s.t. } x_t^P + s_t^P \leq \kappa^P, \qquad (\Psi_t^P) \tag{12}$$

$$0 \leq x_t^P, s_t^P. \qquad (\tilde{\Psi}_t^P, \tilde{\Psi}_t^{PS}) \tag{13}$$

Proposition 2. – *If $\mu_t = 1$, prosumer P utility function is strictly concave in x_t^P, s_t^P and maximized at a single optimum solution of optimization problem (11) under constraints (12), (13).*

– If $\mu_t < 1$, the Nash equilibrium solution of the non-cooperative game is uniquely defined as a parametric function of μ_t, which can be obtained as solution of a fixed point equation.

In both cases, we have the relations: $\Psi_t^P(x_t^{P\star} + s_t^{P\star} - \kappa^P) = 0$, $\tilde{\Psi}_t^P x_t^{P\star} = 0$, $\tilde{\Psi}_t^{PS} s_t^{P\star} = 0$, $\Psi_t^P \geq 0$, $\tilde{\Psi}_t^P \geq 0$, $\tilde{\Psi}_t^{PS} \geq 0$.

Proof. Due to the space limit, the proof has been removed but can be found online at https://hal.archives-ouvertes.fr/hal-01874798v1/document.

Proposition 3. *Suppose that at the optimum $x_t^{P\star} > 0, s_t^{P\star} > 0, \forall P \in \mathcal{P}, \forall t$ and $\boldsymbol{y}_t^\star > 0, \forall t$. There exists a market clearing price upper-bound $\bar{p}$, below which supply-shortages occur on the platform.*

Proof. The proof can be found online at https://hal.archives-ouvertes.fr/hal-01874798v1/document.

Proposition 3 coincides with the results obtained in [3] for Didi Chuxing, the largest ridesharing platform in China: if the platform market clearing price is not high enough, suppliers might lack incentives to share their production on the platform and consumer-shortages might happen. In such cases, optimal design of subsidies might be necessary to give incentives to suppliers (prosumers) to share their supply.

2.3 Modeling of Exchange of Market Platform

In addition to consumers and prosumers, the market platform is assumed to be connected with a surrounding market environment (grid). An example would be the wholesale market on transmission level. It is possible via the exchange to import missing electricity or export excess supply. We assume a simple model for the agent controlling the exchange $q_t \in \mathbb{R}$. The agent optimizes on the price arbitrage between the price of the market platform $p_t^\star$ and the price or cost of electricity at the connected market given by the parameter $c_t^q \in \mathbb{R}$. The price arbitrage is defined by:

$$g(q_t) = p_t^\star q_t - c_t^q q_t. \tag{14}$$

The local MO solves the following optimization problem:

$$\max_{q_t} \ g(q_t). \tag{15}$$

We let q_t be the import (or export) of energy for the community, with perceived revenues $g(q_t)$. We adopt the following convention:

– $q_t \leq 0$ ($\mu_t = 1$), there is an energy lack in the community and the local MO buys the missing quantity to the grid or to another local energy community.
– $q_t > 0$ ($\mu_t < 1$), there is an energy excess in the community and the local MO sells the excess quantity to the grid or to another local energy community.

In order to justify an export of energy ($q_t > 0$), the price on the platform must be at least as high as the price of the exchange ($p_t^\star \geq c_t^q$). The only possible case is that the price of the exchange is below the cost of the prosumers for which they are willing to share part of their generation. As such, the case might also be prevented by the setting of the price floor $\underline{p}$.

2.4   Designing a Trust-Based Mechanism

Consumers and prosumers interact through a sharing platform. The platform operated by the local MO, matches the prosumers and the consumers[3], and sets a clearing price. At each repetition $\nu \in \mathbb{N}^*$ of the game, the platform clearing price profile defined over T_C consecutive time periods, $(p_t^\star)_{t=(\nu-1)T_C+1,...,\nu T_C}$, is determined as solution of an optimal exchange problem [8]. Formally, as will be detailed later on, it is computed as a consensus variable.

After ν repetitions of T_C consecutive time periods, a *reputation* index R_ν^n and an *anciety* index A_ν^n are computed and associated to each node $n \in \mathcal{N}$. Based on these indexes, a miner node is selected in exchange of a mining fee. In practice, nodes in the system compete to solve the complex mathematical program (puzzle) necessary to validate the last block. Reputation indexes are introduced to build decentralized trust-based mechanisms and prevent the emergence of (large-scale) coalitions of nodes with Byzantine behaviors, which would attack the system [5]. In the context of our paper, we can imagine that consumer nodes group together to decrease artificially their aggregated demand (target demands being quite different from one consumer to another, one excess buying from one consumer can compensate a lack of buying from another consumer) and then make the market clearing price decrease, therefore potentially inducing supply-shortages. In Vangulick et al., the reputation index is a linear function of the quantity of energy broadcast previously by the node to mine blocks, the age of the last mined block and a trust index [13]. Furthermore, only consumer nodes can be selected as miners and the selection is made based on a random rule in which the node reputation index is weighted by a random uniform variable. We see no obvious reason to restrict the miner selection to consumer nodes only and define a selection rule less conservative than the one introduced in [13]. Our miner selection rule is based on a fixed-share exponentially-weighted average density function, which is far less energy-greedy than classical PoW methods used in Bitcoin, and less arbitrar than PoS methods used in Ethereum.

The game takes place over νT_C consecutive time periods (or, alternatively, is repeated ν times), taking as input the demand and production schedules of consumers and prosumers. From $(\nu - 1)T_C + 1$ to νT_C, the game timing can be described as follows:

(i) **Miner Selection** A miner node $m_\nu^\star$ is selected based on his reputation index (R_ν^n) and how often he has not been selected in the past, that we will call *anciety* (A_ν^n), following a fixed-share exponentially-weighted average density function, to guarantee the balance between exploration (of new nodes) and exploitation of other nodes having good reputation. Formally, for any $n \in \mathcal{N}$, the probability that node n is selected as a miner is:

---

[3] The matching process itself is out of the scope of the current paper and, as such, will not be detailed here.

$$\mathbb{P}_{n,\nu} := (1-\gamma) \frac{1 - \exp(-\zeta R_\nu^n)}{\underbrace{\sum_{n' \in \mathcal{N}} \left(1 - \exp(-\zeta R_\nu^{n'})\right)}_{\text{exploitation}}} + \gamma \frac{1 - \exp(-\zeta A_\nu^n)}{\underbrace{\sum_{n' \in \mathcal{N}} \left(1 - \exp(-\zeta A_\nu^{n'})\right)}_{\text{exploration}}},$$

$$\forall \nu \in \mathbb{N}^*,$$

$$\mathbb{P}_{n,0} := \frac{1}{N},$$

where $\gamma \geq 0$ is a parameter characterizing the trade-off between exploration and exploitation in the miner selection process and ζ determines the growth rate of the selection probability. The miner node receives a mining fee $\sum_{t=(\nu-1)T_C+1}^{\nu T_C} \Phi_t^{m_\nu^\star}$, and the utility function of the nodes are updated as follows: $\Pi_{C,\nu}^\star := \left[\sum_{t=(\nu-1)T_C+1}^{\nu T_C} \Pi_C(y_t^{C\star}) + \Phi_t^C \mathbf{1}_{m_\nu^\star = C}\right]$, and $\Pi_{P,\nu}^\star := \left[\sum_{t=(\nu-1)T_C+1}^{\nu T_C} \Pi_P(x_t^{P\star}, \boldsymbol{y}_t^\star, \boldsymbol{s}_t^\star) + \Phi_t^P \mathbf{1}_{m_\nu^\star = P}\right].$

(ii) **Consumer-Prosumer Interactions**
  - The local MO computes the clearing price profile $(p_t^\star)_{t=(\nu-1)T_C+1,\dots,\nu T_C}$ and sends it to the consumers and prosumers.
  - Each consumer C computes the demand schedule $\left(y_t^C(p_t^\star)\right)_{t=(\nu-1)T_C+1,\dots,\nu T_C}$ that maximize his utility. Similarly, each prosumer P computes her consumption schedule $\left(x_t^P(p_t^\star)\right)_{t=(\nu-1)T_C+1,\dots,\nu T_C}$ from self-production and the quantity she wants to share on the platform $\left(s_t^P(p_t^\star)\right)_{t=(\nu-1)T_C+1,\dots,\nu T_C}.$

(iii) **Reputation Update** A consensus-based algorithm is run by the platform for a finite positive number n_{iter} of consecutive iterations, reputation index of the nodes are updated based on the divergences between their schedules and algorithm output, leading to the following rule for consumer $C \in \mathcal{C}$ and prosumer $P \in \mathcal{P}$: $R_{\nu+1}^C = R_\nu^C + 1_{\frac{1}{T_C}\sum_{t=1}^{T_C} \|\frac{y_{(\nu-1)T_C+t}^C - y_{(\nu-1)T_C+t}^{C\sharp}}{y_{(\nu-1)T_C+t}^{C\sharp}}\| \leq \tau}$, $R_{\nu+1}^P = R_\nu^P + 1_{\frac{1}{T_C}\sum_{t=1}^{T_C} \|\frac{x_{(\nu-1)T_C+t}^P - x_{(\nu-1)T_C+t}^{P\sharp}}{x_{(\nu-1)T_C+t}^{P\sharp}}\| \leq \tau}$. Variable A_ν^n, called anciety, captures how often node n has *not* been selected as a miner until time period νT_C, it is updated according to the following rule: $A_{\nu+1}^n = A_\nu^n + 1_{m_\nu^\star \neq n}, \forall n \in \mathcal{N}.$

Convergence of the Trust-Based Mechanism. Consider the reputation (resp. anciety) in the node selection probability. We let the gain function $h_\nu(n)$ be the indicator function in the reputation (anciety) update and define the external regret as $\mathcal{R}(\mathbf{m}^\star, n_{iter}) := \sum_{\nu=1}^{n_{iter}} \left(-h(m_\nu^\star)\right) - \min_{n \in \mathcal{N}} \left(-h_\nu(n)\right)$. If the rate is chosen so that $\zeta = \sqrt{\frac{8\ln(N)}{n_{iter}}}$, then $\lim\sup_{n_{iter} \to \infty} \frac{1}{n_{iter}} \mathcal{R}(\mathbf{m}^\star, n_{iter}) = 0$ with probability 1 as well-known in the weighted average forecasting literature.

For n_{iter} large enough, depending on the value of $\gamma \in]0;1]$, regret-minimizing nodes satisfying one or a combination of both criteria will be selected as miners.

Suppose Proposition 3 holds in the rest of the paper. This seems a reasonable assumption, as it is also observed in practice [3]. Assuming that the quantity q_t on the local energy market at each time period is known, the market clearing writes down as a concave optimization problem, that can be interpreted as an optimal exchange problem [1,8]:

$$\max_{y,x,s} \sum_{t=(\nu-1)T_C+1}^{\nu T_C} \left\{ \sum_{C \in \mathcal{C}} \Pi_C(y_t^C) + \sum_{P \in \mathcal{P}} \Pi_P(x_t^P, s_t^P) + g(q_t) \right\}, \qquad (16)$$

$$\text{s.t. } x_t^P + s_t^P \leq \kappa^P, \forall P \in \mathcal{P}, \forall t, \qquad (17)$$

$$0 \leq x_t^P, s_t^P, \forall P \in \mathcal{P}, \forall t, \qquad (18)$$

$$0 \leq y_t^C \leq \kappa^C, \forall C \in \mathcal{C}, \forall t. \qquad (19)$$

3   Solving the Platform Problem Using ADMM

In the platform problem, the local MO determines the clearing price $p_t^\star$ which maximizes the social welfare, under *limited information exchange* between the nodes. To apply decentralized optimization, we reformulate the platform optimization problem (16)–(19) by observing that the objective function can be decomposed onto the consumers and prosumers' decision variables.

3.1   Reformulation as a Consensus Problem

We set $\tilde{g}(q_t) = -c_t^q q_t$. The platform optimization problem can be formulated as follows:

$$\max_{y,x,s} \sum_{t=(\nu-1)T_C+1}^{\nu T_C} \left\{ \sum_{C \in \mathcal{C}} \tilde{U}_C(y_t^C) + \sum_{P \in \mathcal{P}} \left[U_P(x_t^P) - c_P(x_t^P + s_t^P) \right] + \tilde{g}(q_t) \right\}, \quad (20)$$

$$\text{s.t. } \sum_{C \in \mathcal{C}} y_t^C - \sum_{P \in \mathcal{P}} s_t^P + q_t = 0, \forall t, \qquad (p_t^\star) \qquad (21)$$

$$x_t^P + s_t^P \leq \kappa^P, \forall P \in \mathcal{P}, \forall t, \qquad (22)$$

$$0 \leq x_t^P, s_t^P, \forall P \in \mathcal{P}, \forall t, \qquad (23)$$

$$0 \leq y_t^C \leq \kappa^C, \forall C \in \mathcal{C}, \forall t. \qquad (24)$$

The platform problem can be solved in three different ways, that are pictured in Fig. 2:

(a) Joint optimization of (20)–(24). Notice that this requires that the local MO has full information on consumers and prosumers' utilities.

(b) Mixed Complementarity Problem (MCP) reformulation solved as a squared system made of all the first-order stationarity conditions derived in Sect. 2. Notice that his requires that the local MO has full information on consumers and prosumers' preferences.

(c) Iteratively, with limited private information exchanges between the nodes.

We focus on approach *(c)* in this section. To determine the platform clearing price with limited exchange of information between the nodes, we resort to use an algorithmic approach. Our key technical tool is an optimization technique known as the ADMM [1,4,10]. Applying ADMM to our problem, a concave objective function of the form $\sum_{t=(\nu-1)T_C+1}^{\nu T_C} \left(\sum_{P\in\mathcal{P}} f_P(\boldsymbol{X}_t^P) + \sum_{C\in\mathcal{C}} f_C(\boldsymbol{Y}_t^C) \right)$ is maximized subject to some constraints by performing alternating individual optimizations over $f_C(.)$ and $f_P(.)$. While it was originally introduced to achieve faster convergence [1], it was observed in [10] that when the functions $f_C(.)$ and $f_P(.)$ are private information belonging to consumer C and prosumer P, ADMM has the additional advantage of sharing only a small amount of private information between the two parties.

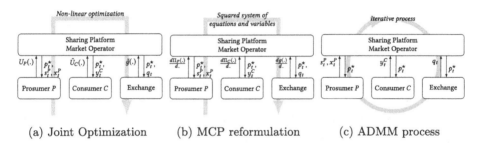

Fig. 2. The three approaches and necessary information exchanges for the local market operator to clear the platform market.

In the general formulation, agents are trying to jointly solve the generic concave optimization problem:

$$\max_{\boldsymbol{X},\boldsymbol{Y},\boldsymbol{Z}} \sum_{t=(\nu-1)T_C+1}^{\nu T_C} \left(\sum_{P\in\mathcal{P}} f_P(\boldsymbol{X}_t^P) + \sum_{C\in\mathcal{C}} f_C(\boldsymbol{Y}_t^C) + h(\boldsymbol{Z}_t) \right), \qquad (25)$$

$$\text{s.t. } \boldsymbol{A}_P \begin{pmatrix} \boldsymbol{X}_t^P \\ \boldsymbol{Z}_t \end{pmatrix} \leq \boldsymbol{b}_P, \forall P \in \mathcal{P}, \forall t,$$

$$\boldsymbol{A}_C \begin{pmatrix} \boldsymbol{Y}_t^C \\ \boldsymbol{Z}_t \end{pmatrix} \leq \boldsymbol{b}_C, \forall C \in \mathcal{C}, \forall t,$$

$$\boldsymbol{B}\boldsymbol{Z}_t \leq \boldsymbol{b}, \forall t.$$

By identification with the relaxed optimal exchange problem (16)–(19), we set:
$\boldsymbol{X}_t^P := (x_t^P, s_t^P)$, $\boldsymbol{Y}_t^C := y_t^C$, and $\boldsymbol{Z}_t := p_t^\star$. Regarding the objective functions,
we set: $f_P(\boldsymbol{X}_t^P) = \Pi_P(x_t^P, s_t^P)$, $f_C(\boldsymbol{Y}_t^C) = \Pi_C(y_t^C)$, $h(\boldsymbol{Z}_t) = g(q_t)$.

The time periods being not linked, we decompose (25) over each time period
t, so that the vectors constraints at t take the form: $\boldsymbol{A}_P = \begin{pmatrix} 1 & 1 & 0 \\ -1 & 0 & 0 \\ 0 & -1 & 0 \end{pmatrix}$, $\boldsymbol{b}_P = \begin{pmatrix} \kappa^P \\ 0 \\ 0 \end{pmatrix}$, $\boldsymbol{A}_C = \begin{pmatrix} 1 & 0 \\ -1 & 0 \end{pmatrix}$, $\boldsymbol{b}_C = \begin{pmatrix} \kappa^C \\ 0 \end{pmatrix}$, and $\boldsymbol{B} = \begin{pmatrix} 1 & -1 \end{pmatrix}$, $\boldsymbol{b} = \begin{pmatrix} \overline{p} & -\underline{p} \end{pmatrix}$.

In our case, $f_n(.)$, $\boldsymbol{b}_n$ constitute the information privately held by node $n \in \mathcal{N}$, whereas $h(.)$, $\boldsymbol{B}$ and $\boldsymbol{b}$ are known to all nodes. We want to solve this concave
optimization problem so that in the optimum, $\boldsymbol{X}^P$ (resp. $\boldsymbol{Y}^C$) is private output
known only by prosumer P (resp. consumer C), but $\boldsymbol{Z}$ may be known to all.
ADMM uses an iterative process to solve the optimization problem and only
shares the iterative updates of the shared variable $\boldsymbol{Z}$.

To make the link with the classical consensus-based approach [1,10], we consider a slightly different concave optimization problem:

$$\max_{u,v} \sum_{t=(\nu-1)T_C+1}^{\nu T_C} \Big(F(\boldsymbol{u}_t) + G(\boldsymbol{v}_t) \Big), \tag{26}$$
$$s.t. \ \boldsymbol{B}_1 \boldsymbol{u}_t \leq \boldsymbol{d}_1, \forall t,$$
$$\boldsymbol{B}_2 \boldsymbol{v}_t \leq \boldsymbol{d}_2, \forall t,$$
$$\boldsymbol{u}_t = \boldsymbol{v}_t, \forall t.$$

Proposition 4. *Problem (26) is a special case of Problem (25).*

Proof of Proposition 4. We can construct Problem (26) where the variables
$\boldsymbol{u}_t$ and $\boldsymbol{v}_t$ both represent an independent copy of $\Big((X_t^P)_{P \in \mathcal{P}}, (y_t^C)_{C \in \mathcal{C}}, Z_t \Big)$. Formally, let the objective function $F(.)$ and $G(.)$ be defined as follows: $F(\boldsymbol{u}_t) = \sum_{P \in \mathcal{P}} f_P(\boldsymbol{X}_t^P) + \frac{h(\boldsymbol{Z}_t)}{2}$, $G(\boldsymbol{v}_t) = \sum_{C \in \mathcal{C}} f_C(\boldsymbol{Y}_t^C) + \frac{h(\boldsymbol{Z}_t)}{2}$. It is easy to see that $\boldsymbol{B}_1$
and $\boldsymbol{d}_1$ can be created by inserting zeros in appropriate places such that the constraint set $\boldsymbol{B}_1 \boldsymbol{u}_t \leq \boldsymbol{d}_1$ reduces to the union of $\boldsymbol{A}_P \Big(X_t^P \ Z_t \Big)^T \leq \boldsymbol{b}_P, \forall P \in \mathcal{P}$ and
$\boldsymbol{B} Z_t \leq \boldsymbol{b}$. $\boldsymbol{B}_2$ and $\boldsymbol{d}_2$ can be generated following the same way, e.g., by inserting
zeros in appropriate places such that the constraint set $\boldsymbol{B}_2 \boldsymbol{v}_t \leq \boldsymbol{d}_2$ reduces to
the union of $\boldsymbol{A}_C (Y_t^C \ Z_t)^T \leq \boldsymbol{b}_C, \forall C \in \mathcal{C}$ and $\boldsymbol{B} Z_t \leq \boldsymbol{b}$. This completes the
construction of Problem (26). $\qquad \square$

3.2   Updating Rules, Privacy Preservation and Stopping Criteria

We define $\mathcal{FS}(\boldsymbol{u}) := \{\boldsymbol{u} | \boldsymbol{B}_1 \boldsymbol{u} \leq \boldsymbol{d}_1\}$ as the feasibility set of $\boldsymbol{u}$ and $\mathcal{FS}(\boldsymbol{v}) := \{\boldsymbol{v} | \boldsymbol{B}_2 \boldsymbol{v} \leq \boldsymbol{d}_2\}$ as the feasibility set of $\boldsymbol{v}$. Optimization problem (26) can be
decomposed in time. So, ADMM solves Problem (26) in an iterative fashion
[10], where for each time period $t \in \{(\nu - 1)T_C + 1, ..., \nu T_C\}$, each iteration k
has three steps as described below:

$$u_t^{k+1} \in \arg \min_{u \in \mathcal{FS}(u)} \left\{ -F(u) + \lambda^{kT} u + \frac{\rho}{2} \|u - v_t^k\|^2 \right\},$$

$$v_t^{k+1} \in \arg \min_{v \in \mathcal{FS}(v)} \left\{ -G(v) + \lambda^{kT} v + \frac{\rho}{2} \|u_t^{k+1} - v\|^2 \right\},$$

$$\lambda_t^{k+1} = \lambda_t^k + \rho \|u_t^{k+1} - v_t^{k+1}\|^2.$$

Note that the consumer and prosumer optimization problems being separable, first step can be solved independently by each prosumer $P \in \mathcal{P}$; second step can be solved independently by each consumer $C \in \mathcal{C}$; while last step is computed by the local MO.

The update steps of the consensus-based ADMM algorithm violate the output privacy requirements because u_t (resp. v_t), which is revealed to the consumers (resp. prosumers), contains a copy of the private output variables of all the other agents. However, the key point to observe is that the optimization problem in each step can be decomposed into components that depend on different individual variables of u_t (resp. v_t). Therefore, the set of components in optimization steps of consumers that depend on the private output of the other agents can effectively be removed from the objective function, and at the same time, the feasible region $\mathcal{FS}(.)$ can be reduced to the feasible region over X_t^P, Z_t for any prosumer P (resp. Y_t^C, Z_t for any consumer C). Hence, the optimization can be carried out in a way that each agent is only revealed her final value X_t^P (Y_t^C) and Z_t. Hence, the output privacy is also preserved by the consensus-based ADMM algorithm.

ADMM iterates satisfy the following:

- objective convergence $\sum_{t=(\nu-1)T_C+1}^{\nu T_C} \left(F(u_t^k) + G(v_t^k) \right) \to_k F^\star + G^\star$, where $F^{\nu,\star} + G^{\nu,\star}$ are the optimal value of the optimization problem (26) with T_C time period look-ahead at ν-th repetition of the platform game.
- dual variable convergence $\lambda^k \to_k p_t^\star$.
- residual convergence $r^k := \|u^k - u^\star\| + \|v^k - v^\star\| \to_k 0$, where $u^\star$, $v^\star$ contain the optimum values for the prosumers and consumers as described in Propositions 1 and 2.

4   Case Study

In this section, an illustrative case study is presented. The objective of the case study is twofold. Firstly, to exhibit the parts of the equilibrium model and the algorithm to solve it (Subsect. 4.1). Secondly, to present the miner selection process and its dynamics (Subsect. 4.2). Note that Subsect. 4.1 focuses on the equilibrium and the impact of consumer preferences with regard to additional characteristics, while Subsect. 4.2 highlights results of the introduced ADMM algorithm on e.g., the number of required iterations, the dynamics of the miner selection process, and the impact of the threshold parameter τ.

In the case study a consumer (C) interacts with two prosumers (P_1 and P_2) via a platform (acting as the intermediate). For his demand, C chooses between two product characteristics: locality (Loc) and RES-based (RES). The preferences of C serve to match his demand with the supply characteristics of P_1 and P_2. Tables 1 and 2 show the values of the parameters to be used throughout the case study, for the consumer and prosumer models, respectively.

Table 1. Input parameter of consumer for case study.

Consumer C			Prosumer 1	Prosumer 2		0	Loc	RES
$y_t^{C\sharp}$	35	α_{Loc}^{CP}	0.1	0.3	ξ^C	-	0.2	0.8
κ^C	80	α_{RES}^{CP}	0.3	1.3	w	1	0.5	0.2

Table 2. Input parameter of prosumers for case study.

Prosumer P	$x_t^{P\sharp}$	κ^P	c_{P2}	c_{P1}	c_{P0}
Prosumer 1	40	60	1	20	0
Prosumer 2	40	60	0.1	10	0

4.1 Impact of Consumer Preferences on Locality and RES-Based Generation

We notice a linear relation between the consumer demand and the market clearing price. This observation coincides with the results that we analytically derived in Proposition 1. We also observe that the market price (and then, the demand) is maximum in case the consumer is indifferent between the two prosumers for his RES-based demand. That is, the location of the prosumer is irrelevant when looking to fulfill his RES preference. Furthermore, preference regarding one prosumer seems to have a limited impact on his demand. We conclude that the definition of the consumer matching preferences for his RES-based demand have a direct impact on his relative consumption, and, on his utility which is maximized in case where the consumer is indifferent between both prosumers. Furthermore, to maximize his utility the consumer can have strong preference regarding RES-based generation but limited preference regarding locality. In general, the consumer should keep his preference regarding locality moderate to maximize his utility.

4.2 Miner Selection: Dynamics and Threshold

In Fig. 3, we represent the error distribution of three nodes, based on market results for 24 market clearings, i.e., for each hour of the day. The error is defined as the normalized difference between the demand (auto-consumption) obtained as outcome of the consensus algorithm and the target value, and used to update the node reputation. Based on these equations, the impact of the error term also depends on the chosen threshold τ. We observe that for values of $\tau > 0.2$, node 2 generates the largest errors. As a result, if $\tau \leq 0.2$, node 2 reputation will

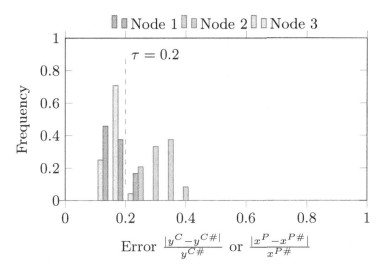

Fig. 3. Frequency of error terms of nodes based on target value (used during update of reputation)

remain unchanged. For $\tau \leq 0.1$ *or* $\tau \geq 0.4$, all the nodes keep their reputation unchanged. Bear in mind that the reputation is updated (i.e., increases by 1) if the node's error is below the threshold τ. Nodes 1 and 3 have lower errors than node 2, with high frequencies for low errors at one node or the other. Meaning that with a $\tau = 0.15$, node 1 has the largest reputation, while a bigger threshold e.g., $\tau = 0.18$ gives node 3 a larger reputation.

The proposed selection process accounts for the exploration and exploitation of miners. Figure 5 illustrates the dynamics of the selection process in terms of γ; parameter that controls for the trade-off between exploration of new miners and the exploitation of miners with high reputation.

If we assume an equal weighting of the exploitation and exploration probabilities, i.e., $\gamma = \frac{1}{2}$ (Fig. 5a), we observe the following:

- node 2's reputation worsened, in respect to the others' nodes, with the number of market clearings. The exploration term is not able to compensate for the low reputation.
- nodes 1 and 3 show a higher selection probability than node 2. This probability oscillates during the first clearings and settles around a value of 0.2. Note that both nodes show values that are close enough to be considered as identical. This, for more than 15 repetitions of the game.

When a higher emphasis is given to the exploration of new miners, that is $\gamma = 0.8$, we observe a quite different picture (Fig. 5b). Node 2 gets selected as a miner in the beginning, and its chance to get selected remains at a relatively high level compared to the others. Forcing the process to look for new miners increases substantially the chances of poor performing nodes to become the miner. This,

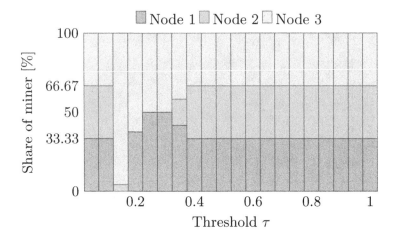

Fig. 4. Miner selection for 24 market clearings with changing values for τ

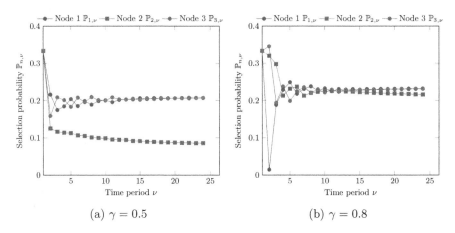

Fig. 5. Dynamics of miner selection process (Parameter γ to control trade-off between exploration and exploitation. $\tau = 0.2$)

even if the reputation is low (and never updates) and the threshold τ is such that with a $\gamma = \frac{1}{2}$ (i.e., no favoritism for exploitation or exploration) this node would never be selected as the miner (see Fig. 4).

As such, a $\gamma = 0.8$ assigns a higher probability to select node 2 as the miner during all clearings than assuming $\gamma = \frac{1}{2}$. Consequently, γ provides a valuable tuning parameter to control the miner selection process.

5   Conclusions

In this paper, we consider consumers and prosumers who interact via a platform. On the one hand, consumers specify their target demand and optimize their demand to the platform in order to find a trade-off between maximizing their usage benefit and minimizing the cost they pay to the platform. On the other hand, prosumers need to determine the amount of generated energy they self-consume and the quantity they share on the platform. Our study introduces product differentiation and consumers's preferences, namely locality and RES-based generation. These preferences are used to match the prosumers generation characteristics. We introduce the probability for a prosumer to be matched to a consumer. In case the consumer demand is larger than the prosumer supply, the matching problem can be decomposed in decoupled optimization problems, that we solve analytically. In case of an excess of supply compared to the demand on the platform, the consumers and prosumers problems remain coupled through the matching probability, giving rise to a non-cooperative game. We provide analytical conditions for the existence and uniqueness of a Nash equilibrium.

We prove the existence of a market clearing price cap below which supply-shortages occur on the platform. Under this assumption (also observed on Didi Chuxing, the largest ridesharing platform in China), we implement ADMM reformulated as a consensus problem, to solve this specific platform issue. A trust-based mechanism is implemented on top of it, to select at each repetition of the game (clearings) the node which validates the transactions (e.g., demand, self-consumption and shared production from the prosumers), that is the node that acts as the miner. The miner node selection is made according to a density function capturing the trade-off between exploitation of nodes with good reputation and exploration of new nodes. The goal is to prevent that the nodes deviate too much from their target schedules, forming coalitions that could work independently of the platform.

Our case study quantifies the impact of consumers' preferences for the matching with prosumers and assess the dynamics of the miner selection process for three nodes. The case study shows that although product differentiation could in theory drive the consumers decision on how to supply his/her demand (based on his/her preferences), in practice the decision's main driver is the price it pays for the product (in this case, energy). In addition, we observed that a tunning parameter that captures the trade-off between exploring for new miners and exploiting nodes with good reputation is relevant for the control of the miner selection process.

References

1. Boyd, S., Parikh, N., Chu, E., Peleato, B., Eckstein, J.: Distributed optimization and statistical learning via the alternating direction method of multipliers. Found. Trends Mach. Learn. **3**(1), 1–122 (2011)
2. Fagnani, F., Zampieri, S.: Randomized consensus algorithms over large scale networks. IEEE J. Sel. Areas Commun. **26**(4), 634–649 (2018)
3. Fang, Z., Huang, L., Wierman, A.: Prices and subsidies in the sharing economy. In: 26th International Conference on World Wide Web - WWW 2017, pp. 53–62. ACM Press (2017). https://doi.org/10.1145/3038912.3052564

4. Höschle, H., Le Cadre, H., Smeers, Y., Papavasiliou, A., Belmans, R.: An ADMM-based method for computing risk-averse equilibrium in capacity markets. IEEE Trans. Power Syst. **33**(5), 4819–4830 (2018)
5. Lamport, L., Shostak, R., Pease, M.: The Byzantine generals problem. ACM Trans. Program. Lang. Syst. **4**(3), 382–401 (1982)
6. Le Cadre, H.: On the efficiency of local electricity markets under decentralized and centralized designs: a multi-leader Stackelberg game analysis. Cent. Eur. J. Oper. Res. (CEJOR) **14**, 1–32 (2018). https://doi.org/10.1007/s10100-018-0521-3. In press
7. Le Cadre, H., Mezghani, I., Papavasiliou, A.: A game-theoretic analysis of transmission-distribution system operator coordination. Eur. J. Oper. Res. (EJOR) **274**(1), 317–339 (2019)
8. Moret, F., Pinson, P.: Energy collectives: a community and fairness based approach to future electricity markets. IEEE Trans. Power Syst. (2018, in press). https://doi.org/10.1109/TPWRS.2018.2808961
9. Munsing, E., Mather, J., Moura, S.: Bolckchains for decentralized optimization of energy resources in microgrid networks. In: International Conference on Control Technology and Applications, IEEE (2017). https://doi.org/10.1109/CCTA.2017.8062773
10. Jiang, A.X., Procaccia, A.D., Qian, Y., Shah, N., Tambe, M.: Defender (mis)coordination in security games via ADMM. In: IJCAI 2013, pp. 220–226. ACM Press (2018)
11. Szabo, N.: Smart contracts: building blocks for digital markets. Extropy **16** (1996)
12. Tushar, W., Yuen, C., Mohsenian-Rad, H., Saha, T., Poor, V., Wood, K.L.: Transforming energy networks via peer to peer energy trading: potential of game theoretic approaches. IEEE Trans. Signal Process. **247**(935), 529–551 (2018)
13. Vangulick, D., Cornélusse, B., Ernst, D.: Blockchain for peer-to-peer energy exchanges: design and recommendations. In: PSCC Conference, Dublin, Ireland (2018)
14. Wang, Y., Saad, W., Han, Z., Poor, V., Başar, B.: A game-theoretic approach to energy trading in the smart grid. IEEE Trans. Smart Grid **5**(935), 1439–1450 (2014)
15. Zhang, C., Wu, J., Long, C., Cheng, M.: Review of existing peer-to-peer energy trading projects. Energy Proc. **105**, 2563–2568 (2016)

Author Index

Printed in the United States
By Bookmasters